T0357210

The Story of Astrophysics in Five Revolutions

The Story of Astrophysics in Five Revolutions

ERSILIA VAUDO

Translated by
VANESSA DI STEFANO

W. W. NORTON & COMPANY
Independent Publishers Since 1923

First published in Italian in 2023 as *Mirabilis: Cinque intuizioni (piú altre in arrivo) che hanno rivoluzionato la nostra idea di universo.*

Printed in the United States of America
First Edition

Manufacturing by Versa Press
Book design by Chrissy Kurpeski
Production manager: Anna Oler

ISBN: 978-1-324-08927-8

W. W. Norton & Company, Inc.
500 Fifth Avenue, New York, NY 10110
www.wwnorton.com

W. W. Norton & Company Ltd.
15 Carlisle Street, London W1D 3BS

1 2 3 4 5 6 7 8 9 0

To Francesco and Caterina

Contents

The Story of Astrophysics in Five Revolutions

A Jolt of Reality

There once was a hidden side of the Moon that no one had ever set foot on. But not anymore.

On January 3, 2019, "Jade Rabbit"—the rover *Yutu-2* from the Chinese *Chang'e 4* mission—broke the spell by alighting on and exploring that remote, gray world up close. *The dark side of the moon.*

In primary school, they had explained to us that because of a sophisticated cosmic synchrony, as the Moon revolves around Earth it also revolves on its own axis. So no matter where we are in the world, if we look up at the Moon what we see is always and only the same half. At the time, the mechanics of it were beyond me, but that cold, structured precision immediately seemed intentional. The ruse of an inaccessible world, a magical and private place concealed behind a permanent façade that is at times dark and at others radiant.

That is how it began, with the arrival of *Yutu-2* and the loss of a mystery that I still miss, in the year that marked the 50th anniversary of the landing on the Moon.

On July 21, 1969, Neil Armstrong became the first man to step onto our only satellite. Moments later he was joined by Buzz Aldrin, and together they strolled along the Sea of Tranquility for a short, or perhaps very long, time in that alien world. Two hours, 31 minutes, and 40 seconds.

When Armstrong set his foot down in the lunar dust, Earth held its breath. Six hundred million people, one-fifth of the human race at the time, stood transfixed in front of those black-and-white images, striving to understand the magnitude, or even the reality, of that extraordinary moment.

Crossing the abstract frontier between us and other worlds was a momentous step that triggered an irreversible change for the destiny of human presence in the cosmos. Yet, that first footstep and that first stroll were such a unique achievement, and the emotion of it so intense, that the moment remains crystallized in time. Aside from Armstrong and Aldrin, few people know the names of the other astronauts who have been to the Moon. How many there have been in total. Who the last footprint belongs to.

Since Eugene Cernan stepped off the Moon's surface in 1972, no one has returned. What seemed like the dawn of a journey for humanity turned out not to be. Instead, what changed forever was the scope of the possibilities open to us.

* * *

Speaking of the first man on the Moon, Jean-Jacques Dordain—who was director general of the European Space Agency (ESA) for a long time and is one of the most charismatic and influential personalities in the international space sector—describes his first encounter with Armstrong as follows. Paris Air Show 1971, Le Bourget. Neil Armstrong, by now an undisputed global icon, arrives and is immediately swarmed by a huge crowd. The young engineer Dordain pushes his way through. When Armstrong finds Dordain standing before him, breathless and excited, he offers him his hand.

"I am honored to shake your hand," says the intrepid engineer. "But I'd like to ask you a favor."

"How can I help you?" replies Armstrong, surprised.

"Can I stand on your feet?"

"On my feet, why?"

"Well, so that I can say that I stood on the feet that stood on the Moon."

And that is what he did.

* * *

Just a few months before the landing on the Moon, something equally exceptional had happened. It was Christmas Eve 1968. Frank Borman, Jim Lovell, and Bill Anders of the *Apollo 8* mission, the first men to leave Earth and orbit around our satellite, were also the first to see its hidden side with their own eyes. Over the course of about 20 hours, the crew circled the Moon an impressive 10 times. They were reading passages from Genesis during a televised transmission to Earth that would be etched into history, when, on the fourth lap, an unexpected image left them speechless. It was a sight for which a new word would be invented: *Earthrise*. The rising of Earth over the Moon's horizon. Bill Anders grabbed his Hasselblad and shot what in years to come would be considered one of the photos that changed the world. An extraordinarily beautiful sight that no eye had ever—ever, ever, ever—gazed upon before. Earth hovering there, suspended in the darkness of a cosmic night, wrapped in a thin blue halo, an imperceptible buffer between our world and a void with no breath.

The lesson is clear. Do not even imagine that you matter. You are fragile, tiny, a mere speck in an indifferent Universe. But you have come this far. And it was probably this sight, before that first footstep in the lunar dust, that gave new meaning to our presence in the cosmos.

"Why is there something rather than nothing?" asked Gottfried Wilhelm Leibniz in his *Principles of Nature and of Grace Founded in Reason,* adding that "nothing" would have been a much simpler solution after all. A complex philosophical question, and the most charged according to Umberto Eco. Asking questions is one of the conditions of "being," the inexorable tension toward what we do not know. The word *desire* comes from the Latin *de-*, "negation," and *sidus,* "star." To be far from the stars. This distance from the stars can be understood in many different ways. It could be the epitome of an absence, a cosmic indifference, with the Universe and its distant stars simply not caring about us. Or a lack of good omens. Or something else. Desire as in "missing stars," a longing for reconnection, an inner yearning for what disregards us but to which we nevertheless belong. A distance, a desire, that has plagued us forever.

* * *

Today we have two outposts in the cosmos. The probes *Voyager 1* and *Voyager 2* began their journey in 1977 and have been hurtling across interstellar space ever since, skirting the edges of human presence in the Universe.

Riding on the slingshot effect created in this case by a rare alignment of the giant planets—which occurs only once every 175 years—the two spacecrafts managed to traverse the entire Solar System and, after decades, breach its frontier and enter a dark and very cold world of gas, dust, and cosmic rays that shoot out at the speed of light. *Voyager 1* got there first, in August 2012, and *Voyager 2* joined it in November 2018. They are still traveling to imagined destinations and, like the Wise Men, they are not traveling empty-handed. Each is carrying a Golden Record, a gold-plated copper disc that is supposed to explain who we are to whoever finds it.

The choice of content for the Golden Record was in a sense the first curated work intended for an interstellar audience. NASA entrusted this task to a commission chaired by Carl Sagan, a renowned astrophysicist and professor at Cornell University and a key character in this very unique "first."

There was plenty of debate between the members of the commission, as was inevitable considering the extraordinary circumstances: for the first time, in anticipation of an encounter with unknown interlocutors, an attempt was being made to put together a concise and effective account of the planet and who we are. The Golden Record ended up containing 115 images, among them a snowflake, a photo of a family, the Tāj Maḥal, and a page from Isaac Newton's *Principia;* as well as audio files that included "good morning" and "good evening" spoken in 55 different languages, the "sounds of Earth," such as the formation of an avalanche, the trumpeting of an elephant, and the sound of a kiss, and about 90 minutes of music from various eras. And finally, a lot of math, too: graphics explaining numbers 1 to 10, basic calculations, fractions and multiplication, and even a conversion table of units of measurement.

In the years after the launch of the two spacecrafts, Sagan apparently repeatedly asked for a final photo of Earth to be taken, but the idea was systematically rejected, not least because the image would have been devoid of scientific value. He did not give up, however. It was Valentine's Day, 1990. As *Voyager 1* sped toward the edge of the Solar System, a stubborn Sagan managed to convince NASA's administrator, Richard Truly: as it was about to leave the Solar System, *Voyager 1* turned around and looked back one last time. The photograph it took, from a distance of 6 billion kilometers, shows a blurred, remote speck, a "Pale Blue Dot," as Sagan himself called it in a famous 1994 lecture at Cornell University: "a mote of dust suspended in a sunbeam.... The only home we've ever known."

Seeing an unexpected image for the first time is always moving. In that extraordinary moment when we go from ignorance to awareness, a new perspective emerges, and we are no longer the same.

"Il nous faut à chaque instant la secousse du réel" (At every instant we need a jolt of reality), wrote Victor Hugo in *Promontorium Somnii*, after the astronomer François Arago had given Hugo his first view of the Moon through a telescope at the Paris Observatory.

It was a clear summer evening in 1834. With his eye trained on the telescope, Hugo at first failed to see anything interesting, but then something happened: "Visibility improved. . . . Everything, however, remained indistinct, and there was little change except from pale to dark. . . . The sense of depth and loss of reality was terrible. Yet, reality was there." Hugo shuddered. "You find yourself face to face [with the Moon] within the shadows, with this map of the Unknown. The effect is terrifying. The inaccessible is almost touchable. The invisible seen."

That step onto the Moon, the rising of Earth, a greeting from the edge of the Solar System. Jolts of reality, points of no return in that nebulous idea of the world that human experience grants us.

Still in France, years before Hugo "met" the Moon, gazing skyward with noses in the air was joined by the prospect of looking down upon the world below. One could *fly*. This opened a whole new and much-desired dimension that would otherwise have been inaccessible to us. The thrill of height, of hovering, of that silence.

In addition to the anecdote about Hugo and the early excitement of flying in France, Julian Barnes recounts in his wonderful *Levels of Life* what the physicist Jacques Charles, who was the first to make an ascent in a hydrogen balloon, said about the experience: "'When I felt myself escaping from the Earth, [. . .] my reaction was not pleasure but *happiness*.' It was 'a moral feeling', he added. 'I could *hear myself living*, so to speak.'"

CHAPTER 1

The Unbearable Lightness of Gravity

Apples Fall, Comets Spin, the Genius Sees

Believe it or not, sometimes we can see into the future. For example, we can predict what starry nights will look like for the next 1.6 million years. This wonder is thanks to the remarkable results obtained by ESA's Gaia mission, which measured the position and velocity of some 2 billion stars populating our galaxy. A future that is predetermined and deterministic, far from the notion of a slippery and elusive time, but which remains unknown.

The capacity to predict the future has certainly played a crucial role in our evolution. Being able to anticipate the attack of a predator or the changing of the seasons has enabled us to survive and reproduce. Neuroscientist Dean Buonomano explains this well in his book *Your Brain is a Time Machine.* Our brain has to accumulate an enormous amount of information to anticipate what lies ahead. It is, in essence, a device that remembers the past in order to predict the future; a machine that, while it may not understand the nature of time, can nonetheless make sense of it. We do not, in fact, have a sensory organ dedicated to "feeling" time. Yet we perceive its passage to the same extent as we perceive the color of objects. The temporal asymmetry between cause and effect, for example, is deeply encoded in our minds, even down to the tiny difference between the arrival of a sound in the right ear and its arrival in the left. And, because we do not have an organ specifically dedicated to time, to understand it we effectively co-opt the neural circuits we use to map space.

Using space to describe the flow of time: a universal instinct. Such was the finding of the research carried out by cognitive scientists Kensy

Cooperrider and Rafael Núñez and presented in an article in *Scientific American* titled "How We Make Sense of Time." Essentially, all cultures use spatial metaphors to talk about time: in order to be able to deal with the concept of time, human beings need to resort to their understanding and experience of physical space, relying on reference models that relate time to space. Time duration is expressed in terms of dimensions—a short weekend, a long wait—while the passing of time is defined in terms of movement: time flies, moving on with one's life. What changes across cultures is the collocation of the past, present, and future. In cultures where the script goes from left to right, the sequence of temporal events follows the same direction. The past is behind, the present here, and the future ahead. By contrast, other cultures—such as that of the Aymara in the Andes or of the Yopno in Papua New Guinea, which lack a written reference—visualize time in a sequence inverted 180 degrees: the past is in front of us, we can see it because we know it, and thus discuss it; whereas the future is unknown to us, the subject of speculation, and hence lies behind us, where our gaze cannot reach.

As outlandish as it may seem, this latter representation of time is consistent with how the Universe is presented to us. When we look at the sky, the past is in front of us.

The Sun, stars, and galaxies that we see are actually projections of a time gone by. Almost like magic. The fact that light travels at its own speed—therefore not instantaneously—gives us the superpower to look into the past, even without any devices or telescopes. As we drink an aperitif and watch the Sun go down with a friend, the beautiful red sphere about to disappear behind the horizon in that shared present has been gone for more than 8 minutes. Realizing this is deeply stirring. The remote lights that pepper the night sky are glimmering flashes of times past twinkling in the silence. The beautiful star Rigel, shining brightly to the naked eye in the constellation Orion, is an image, a "photograph" that reaches us today but which was "taken" around the time Dante Alighieri was born. A starry sky. Many layers of time overlapping, all within a single darkness.

Aristotle was one of the very first to raise the question of time, equating it to a measure of change. In the fourth book of *Physics*, he wrote that time is the "number of motion in respect of 'before' and 'after,'" contingent on the existence of perceptible reality. There is, therefore, a very close link between time and the sublunary world.

Isaac Newton was the first person to give a precise definition of what time is and, through that, what space is. The equations of motion he was working on required it. And he defined them in an absolute sense, without any dependence on or potential influence from anything else. According to Newton, space and time are like two empty vessels within which all of reality can be slotted. Space is therefore a static, immutable place within which celestial objects move. And time follows a universal rhythm, which applies to everyone, everywhere. Newton's time is absolute, true, mathematical. His three laws of motion, which are based on these concepts of absolute time and space, explain the dynamics of objects large enough for direct observation. If time and space are absolute, nothing can influence them. Space is not conditioned by what it contains, time flows without being bound by space, and both are independent of any observer.

Bodies move within this absolute space, and absolute time is what is required for the laws of physics to be the same at every instant. Newton wrote that "all motions can be accelerated and retarded, but the flow of absolute time cannot be changed." And this, in short, is also our perception as "Newtonian earthlings," the intuitive impression we have of how the world works. It is not until more than 200 years after Newton that we realize our perception of what is happening around us is limited, that our senses do not stretch beyond the edges of a small portion of reality. With the emergence of quantum mechanics and the special and general theories of relativity, we have come to realize that the Universe works in ways of which we are largely unaware.

According to Newton's equations of motion, a body begins to move if it is subjected to a force that accelerates it. And, if there is an acceleration, there must be a force acting on that body. This force causes an apple falling from a tree to move through space and eventually hit the ground. However, according to Newton, there is no reason why this force cannot also extend beyond the top of the tree. Taking his thinking further, he concluded that the movement of the planets up there in the sky could also be due to the same cause. If a celestial body moves along its orbit, it does so as a result of the same physical law that dictates how an apple falls from a tree.

A falling apple and a planet stuck in its eternal orbit. An unimaginable parallel until then. This extraordinary insight marked an irreversible turning point by bringing together dynamics that seemed to belong to two very different worlds, the celestial and the terrestrial. From the time of Aristotle, it was believed that there was a clear separation between celestial bodies and terrestrial ones, to the point that the former were composed of a specific element (ether) that could not be found in the sublunary world and would therefore follow different laws, moving in a uniform circular motion due to a total absence of friction. Even Copernicus, to whom we owe the formulation of the heliocentric theory, had not discussed the mechanism that governed the motion of the planets.

This fundamental shift in astronomy began thanks to the work of Galileo Galilei and Johannes Kepler. Giacomo Leopardi's deep admiration for the German scientist Kepler was clear in his *Storia dell'Astronomia*:

> Kepler was Newton's forerunner. Nature, which had done so much for [Newton], lay fallow in order to elevate the English philos-

> opher. But had he not been preceded by Galilei and Kepler, he would have had to do what they had done, and his knowledge would not have reached the sublime level that it did. [Kepler] was a great man, a remarkable man who deserved the distinguished title of Father of Astronomy.

In his book *Astronomia Nova,* published in 1609, Kepler used Copernican cosmology, Tycho Brahe's more detailed observations, and, in particular, Galileo's experimental method to formulate the two laws that mathematically explain the motion of the planets around the Sun on the basis of elliptical orbits. Ten years later, in *The Harmony of the World,* he added a third law, establishing a precise mathematical relationship between the distance of a planet from the Sun and the time it takes the planet to complete one full revolution. This brought to an end a very intense study that had occupied him for 22 years. He was the first to realize that the Sun exerted a force on the planets that bound them to their orbits, even though he did not investigate the nature of that force.

Nevertheless, Kepler and the formulation of his laws were crucial in the development of a new, revolutionary astronomy. He was, in fact, as Maria Popova writes in *Figuring,* the first astronomer to propose a scientific method for predicting eclipses, the first to link mathematical astronomy to material reality, and the first to demonstrate that physical forces move celestial bodies in a computable manner. All this while he wrote horoscopes for court nobles, scrambled to save his mother from accusations of witchcraft, and continued to defend the idea that Earth was a huge, sentient, disease-prone, animate body, which sometimes even suffered from digestive disorders.

Popova goes on to say that Kepler was baffled by the hypothesis that the Universe was steered by some kind of watchmaker, and a divine one at that. He preferred to focus on the idea of "a single magnetic force" that somehow set the celestial machine in motion, "not something like a divine organism, but rather something like a clockwork in which a single weight drives all the gears." This was not so far removed from Newton's concept, and, until then, no one had found a way to explain the motion of the planets without resorting to some notion of divinity.

When that legendary apple fell from its famous

tree, Newton had been pondering just what force Kepler had been envisioning. The second law of motion developed by Newton states that if a body, either stationary or moving in uniform rectilinear motion, changes its state, then a force, represented by the equation $F = ma$, must have intervened. In other words, the acceleration (a) of a body is proportional to the force (F) applied to it, whose direction is the same, and is inversely proportional to the mass (m) of the body.

If the apple attached to the tree falls, it means that it has been subjected to acceleration, and therefore some force must have acted upon it. Aware of Galileo's work on the motion of projectiles falling to the ground in a curved trajectory, and reflecting on Kepler's third law as applied to the movement of the Moon around Earth, Newton developed the idea of a force "extending across the Moon's sphere, all the way to the Earth" without "any dips or boundaries."

Newton's extraordinary vision was to recognize the same underlying cause behind occurrences that had hitherto been thought to be on different planes of reality: that behind the eternal movement of distant planets and the movement of a fruit falling off a tree lies the same invisible hand. Thus the first of the four fundamental forces known today entered the scene, the king of them all: gravity.

According to Newton's law of universal gravitation, the force by which two bodies attract each other is proportional to their masses and inversely proportional to the square of the distance between them: the greater the mass of the bodies, the greater the force of attraction. Whereas the reverse happens with distance: the force becomes weaker as the distance between the bodies increases—similar to seeing the light of a lighthouse lose its intensity as we move farther away from it. In 1687, Newton formulated this law, together with the three principles of motion, in *Philosophiæ Naturalis Principia Mathematica,* known more simply as the *Principia.* A link was thus established between falling apples and spinning comets.

Newton's insight, bringing together seemingly disparate phenomena under a single cause, was in essence an achievement in simplicity. Newton heralded a decisive turning point in scientific perspective, establishing the possibility of a more complete and broader understanding of the way the Universe operates, unfolding beyond the narrow and deceptive perimeter of human experience and perception.

Those future and deterministic night skies charted by the Gaia mission, the prediction of the movements of the stars, the changing shapes of the constellations, all stem from what Newton gave us. The law of universal gravitation, an elegant mathematical equation, together with the laws of motion, describe a large proportion of celestial movement quite accurately. And they do it so well that we still rely on them today to launch probes and satellites into space, to land on other planets, or to fly toward a comet.

* * *

The story of Newton's apple is well known, striking him as it fell from the very tree in his mother's garden in the shade of which the great physicist had sat pondering the motion of the planets. This story was first floated by François Marie Arouet, otherwise known as Voltaire, in the unexpected context of an essay on epic poetry. A few months after attending Newton's ceremonious funeral at Westminster in 1727, the French philosopher, still in his forties, went to visit the scientist's niece,

Catherine Barton Conduit. During their conversation, she told Voltaire about the tree, and the French writer, fearing that the anecdote would be stolen from him by some gazetteer, romanticized it a little and quickly included it as an off-topic digression in the work he was publishing at the time, where it was destined to become immortal.

> According to the story told to me by his niece (Mme Conduit), one day in 1666, having retreated to the countryside and seeing some fruit falling from a tree, Newton fell into a deep meditation about what causes all bodies to be propelled in this way along a straight line, which, if it were extended, would pass more or less through the center of the Earth.

Fortunately, by the time the story was published, Robert Hooke had been long dead. The English scientist, and Newton's historic rival, had claimed to be the first to have had the insight that Newton would later formalize in his law of universal gravitation. Hooke had measured the oscillation of a pendulum and hypothesized that the motion of planets could also be calculated on the assumption of an attraction inversely proportional to the square of the distance from the source. The alleged rivalry between Newton and Hooke is legendary, and it continued even after Hooke's death. When Newton became president of the Royal Society, he not only used his influence to sweep his opponent's work under the carpet, but apparently even had his portraits destroyed. At a time in history when the boundaries between alchemy and chemistry were rather blurred, Newton also devoted himself—with great discretion—to alchemy, most notably in the latter stages of his life. John Maynard Keynes, who acquired many of Newton's writings on the subject, wrote in a speech for the Royal Society: "Newton was not the first of the age of reason. He was the last of the magicians."

Credible or not, Voltaire's version is the one that stuck, making the tree in Newton's mother's garden—which is still there, on the family estate in Lincolnshire—the embodiment of one of the most innovative insights ever gained by a human being. So much so that, 300 years later, NASA took a small piece of it to the International Space Station and created a somewhat disorienting phenomenon: the wood of the tree that had inspired Newton's theory of universal gravitation suddenly found itself floating in space, as if in the absence of gravity.

* * *

Decades were to pass before the law of universal gravitation was accepted in continental Europe, where Descartes's theory, a conjectural hypothesis involving singular vortices, was still widely accepted. It came about thanks to the scientific "propaganda" carried out by two outstanding minds of 18th-century France. One was Voltaire himself, who wrote *The Elements of Sir Isaac Newton's Philosophy* in a popularized form. The other was the Marquise Émilie du Châtelet, a liberated woman who was extremely cultured and passionate about science. A great friend of Voltaire's, she gave him refuge at her château in Cirey, close to the border with Lorraine, when the philosopher was threatened with arrest for his *Philosophical Letters*, which were considered too subversive. They lived together there for many years and established the first private physics laboratory, full of all kinds of scientific instruments, where they conducted experiments on the propagation of light, the nature of fire, and, later on, kinetic energy. To her we owe the first contributions on infrared radiation and the theorization of the principle of the conservation of energy, as well as her *Dissertation on the Nature and Spread of Fire*, the first written by a woman to be published by the French Academy of Sciences. She was also the first to have translated Newton's *Principia* into French, which is still a reference text today.

It was an extraordinary breakthrough to realize that Earth, Neptune, and Mercury glide along their orbits, in broad, lazy circles, for the same reason that apples fall from trees. However, one may naturally wonder why, if the hand of gravity is always acting to attract and pull downwards, the Earth, Neptune, and Mercury do not fall onto the Sun instead of revolving around it.

Let us do a thought experiment. When an apple drops from a tree, it falls down in a vertical line. If, however, we launch a cannonball horizontally, it will follow a straight trajectory for a while, then curve and finally strike the ground. The greater the speed at which the cannonball is launched, the farther the point at which it will drop to the ground. Therefore, the object's initial position and velocity affect the trajectory of its movement.

Newton imagined a cannonball being launched horizontally at such a high speed that it would go all the way around the planet and return to where it started, without ever touching the ground. If this happened, the curvature of its trajectory would correspond precisely to the curvature of Earth. The cannonball would thus complete an entire orbit, an expression of the perfect balance between the forward motion of a body that, once launched, continues moving in a straight line (the inertia of the cannonball) and the gravitational attraction of a much larger body (Earth) trying to pull it down. This is how, 4.5 billion years ago, when the Solar System was formed from an enormous cloud of swirling gas and dust, Earth, Neptune, Mercury, and the other planets were trapped by the Sun's gravitational force and locked, at different distances according to their initial positions and speeds, into a perpetual loop.

The difference between the motion of the apple falling from the tree and the planets revolving around the Sun by the force of gravity is only a question of speed, initial positions, and the mass attracting them. In essence, both the apple and the planets are doing the same thing. They are falling. Traveling along an orbit is equivalent to experiencing the thrill of free fall.

* * *

For more than 20 years, there has been a human outpost in space. It is the International Space Station, or ISS. In those cylindrical pods, crammed with wires and mysterious objects, we have often seen astronauts moving about as if they were floating. Screwdrivers chasing each other, long hair fluttering, drops of water hovering weightlessly. How wonderful, to see them free from the "tyranny"—we all wish we could fly—of gravity that keeps our feet on the ground. You might think that this is nothing more than the effect of being far away, of being "in space." But that is not true. The International Space Station is about 400 kilometers, or 254 miles, above Earth, and the force of gravity is very much present at that altitude. It is only slightly weaker, by about 10 percent, than what we "feel" at sea level. An individual that weighs 100 pounds on the ground floor of an imaginary 254-mile-high skyscraper would weigh 90 pounds—10 percent less—if that person stepped onto a scale at the top of the skyscraper. The mystery of that weightless world within the ISS is therefore due not to being in space but rather to how one moves about within space. The ISS is orbiting Earth at a speed of about 17,500 miles per hour and completes a full circle every 90 minutes or so. This means in a 24-hour period, the astronauts enjoy the beauty of the sunrise and sunset about 15 to 16 times. The driving force behind these spinning orbits, as Newton has taught us, is the force of gravity.

The lightweight and airborne visitors at the International Space Station are not, therefore, flying. Instead, just like the apple from the tree or a diver off a cliff, they are doing the opposite. They are falling.

Gravity, whose influence is infinite, is also the most extraordinary of designers: a meticulous and implacable hand, hidden in the spherical perfection of celestial bodies that glow, float in the dark, and transform into something else. The larger a body is, the more gravity will shape it, smooth out its curves, and give it a sublime sphericity.

Mars, for example, is smaller than our planet, and therefore also lumpier. Its gravity, being approximately one-third that of Earth, cannot completely level out its protrusions: the highest Martian peak, the volcano Olympus Mons, measures a good 16 miles and is the highest in the Solar System.

The weaker the gravity, the greater the likelihood of ending up with bizarre shapes. Asteroids, comets, and meteorites are cosmic objects that are small, imperfect, and differ from each other, spinning and chasing each other around the Solar System, rebelling, through their quirks, against the elegant roundness of the celestial spheres.

On one of these racing bodies, with a bizarre shape and a surface as soft as the foam of a cappuccino, some 300 million miles from Earth, we were lucky enough to land one day.

In August 2014, after a 10-year voyage, ESA's *Rosetta* probe arrived for a celestial blind date with a beautiful comet—the term *comet* comes from the Greek word for "long hair"—that has the unpronounceable name of 67P/Churyumov-Gerasimenko, is made of ice and gas, measures around 2.5 miles in diameter, and has a rather unusual shape: a sort of small duck.

To get there, *Rosetta* traveled more than 4 billion miles by closely skirting the planets, including Earth three times. After each flyby of a planet, the probe traveled a full orbit around the Sun (five in all) to charge itself with energy, using the gravitational field of the celestial bodies in a sort of slingshot effect to increase its speed. Along its journey, it also came into close contact with two asteroids, Šteins and Lutetia, and when it was still 3 years away from reaching the comet, *Rosetta* put its instruments to "sleep" in order to save energy. On its silent trajectory toward that distant rendezvous, *Rosetta* thus spent no less than 31 months in hibernation (a feat never achieved before!), with all instruments deactivated except for four internal clocks, which, without any intervention from Earth, sounded the alarm on January 20, 2014. It was about to reach its destination. A few months later, *Rosetta* finally made contact with the little comet. Through a series of incredibly complex approach maneuvers, the spacecraft positioned itself in such a way that it rotated around its target in a smooth minuet with which it accompanied the comet for a long time.

One of the objectives of the mission was to understand the mystery of the presence of water on our planet, which, as one of several possible scenarios, may have arrived by way of these celestial bodies.

It was November 12, 2014, when *Rosetta*, now close enough to the comet to feel its, albeit very weak, gravity, dropped *Philae*: a little robot the size of a washing machine, which fell toward its destination almost in slow motion. The 13-mile descent took about 7 hours before *Philae* managed to land on that pebble on its way to the Sun. The image taken by *Rosetta* of the little *Philae* gliding in the dark is itself exciting. The enthusiasm when the signal was received was enormous. The landing of the small robot, after a few hiccups—it bounced three times before finally coming to rest on that frozen core—heralded an unprecedented success. Europe on a comet. The imprint of our robotic foot on the surface of a celestial body that no one had ever ventured onto before, very far from here, in a Universe that has seemed just a little less infinite since that day.

Gravity pulled us together, allowed us to gently land on the comet, and to travel together for a part of the journey toward the Sun. The mission yielded many scientific results, but the mystery of the water still remains. The water on Comet 67P was found to be different from what fills our oceans. Of course, it is still the very familiar H_2O formula, but rendered "heavier" by the rich presence of deuterium, an isotope of hydrogen. And so, how the water we know made its way to Earth still remains to be discovered.

* * *

When you first come across Newton's formula for universal gravitation, you inevitably wonder why it is impossible to see the pull of attraction between the objects around you; for example, why does the coffee cup not collide with the coffee pot?

The reason is simple: no matter how difficult it is to carry a refrigerator on your back to your new sixth-floor apartment or how much energy it takes to launch a rocket into space, gravity is actually a very weak force.

In this world of electrons, protons, and other particles, there are only four forces that account for the fundamental interactions. The first two are called, rather unimaginatively, strong nuclear force and weak nuclear force: they govern physical processes in the microscopic world and have a range of influence confined to atomic distances. Which is why we do not often hear about them unless we work in a nuclear power plant or a physics department. By contrast, we are quite familiar with the other two forces: gravity and electromagnetism, both of which have an infinite range of action.

It might be hard to believe that all possible interactions are the result of just four forces. It feels like we encounter many others in our everyday experience: the friction that allows a car to brake, the buoyancy that keeps us afloat while swimming, the tension of a bow ready to shoot an arrow. They seem like different forces, and yet, when analyzed in detail, they can all be traced back to the four fundamental forces.

Very often it is the electromagnetic one. For example, when the friction between two surfaces, which is caused by the contact between them and which allows us to walk or write, is analyzed under a microscope, it turns out to be the combined effect of the electromagnetic interaction between the atoms of one surface and those of the other. Thus, there is a microscopic reality that, if delved into, explains phenomena that would otherwise be impossible for us to perceive.

Gravity occupies last place on the intensity scale in a list that sees strong nuclear force at the top, followed by electromagnetic force, and then weak nuclear force. What is astonishing is how far down the list that last place is: very, very far from the top three positions.

Let us use an example to understand the gap in intensity between electromagnetic force and the force of gravity. Stepping back into the microscopic world, we will consider one of the simplest physical systems: the hydrogen atom. A single electron and a nucleus consisting of a single proton, both with an electric charge and mass. If we calculate the electromagnetic force and the gravitational force acting on the particles, we find that the former is more intense than the latter by a factor of 10^{39}, that is, 10 followed by 38 zeros, or in other words, a hundred trillion trillion trillion. Thus, at the atomic level, the force of gravity has practically no effect.

Instead, it dominates when distances increase to a cosmic scale. We will never see any discernible attraction between the coffee pot and the cup because the force of gravity is so weak. Other forces, such as friction, are much more significant and end up obscuring the existence of gravity from the naked eye.

This difference between gravity and the other forces—that is, the reason for its inherent weakness—is one of the great, as yet unsolved, mysteries of physics. Physicists call it "the hierarchy problem," and it may have profound implications that still elude us. This mystery might have something to do with the Higgs boson, or with the existence of a multitude of universes, or perhaps with other equally fascinating explanations.

Here is one of them. Let us suppose that our Universe has many more spatial dimensions than we are aware of, as envisioned by some theories of physics, such as string theory. So not only those related to height, width, and length but also others, which are well hidden. In such a universe, rich in new spatial dimensions, we can imagine gravity as a force that has a similar intensity to the other three forces, but in the three-dimensional-space world we experience, it appears weak because it is diluted across a multiplicity of available dimensions, both visible and hidden.

* * *

A very large number of protons must agglomerate to ensure that, in the tug-of-war between the gravitational force that attracts and the electromagnetic force that repels, gravity asserts itself and a star can be born.

Nature uses a trick that prevents electric repulsion between protons. In fact, in the interstellar cloud—the mother of the star, and so a kind of grandmother to us—the vast majority of protons are in the form of hydrogen atoms (one proton and one electron) and therefore have zero overall charge. Under these conditions, there is no electric repulsive force, and the atoms are attracted to each other because of weak gravitational force. Whether or not a star then emerges from that cloud depends on various factors.

The first is that the thermal agitation of the cloud, which tends to break up the structure, must be minor. In other words, the cloud must be very cold (a few dozen degrees above absolute zero). Under these conditions, gravity can do its thing as an organizational force undisturbed. Any small concentration of gas begins to attract the surrounding mass and triggers a process of aggregation that leads to a concentration of many, many atoms. Whether this becomes a star (like the Sun) or a planet (like Jupiter) depends on how much mass there is in the cloud and its composition. Under certain conditions, as it grows increasingly dense, the cloud becomes opaque, and the gravitational energy that is released in this process cannot be radiated away. The core of the cloud then gets hotter and hotter, and when it exceeds about 10 million degrees Celsius, the first thermonuclear fusion reactions are triggered, the ones that make it shine, and a star is born. If there is less mass, nothing is triggered. The minimum mass of celestial bodies that have become stars is about 0.08 times that of the Sun.

Therefore, not all gaseous bodies are able to light up in the dark. This is, perhaps, the biggest regret of the largest planet in our Solar System, Jupiter, which "narrowly" missed out on becoming a shining star.

Gigantic and gaseous, Jupiter is mainly composed of hydrogen, like most stars; its sheer volume is the equivalent to 1,321 Earths, and its mass is two and a half times that of all the other planets in the Solar System put together. Those are pretty impressive numbers; yet, its mass is only a thousandth that of the Sun. During its formation, Jupiter thus lacked the mass necessary for its internal pressure to overcome the repulsion of atoms and set off the thermonuclear reactions that would have made it shine in the dark, drunk on its own light.

If it had been a little more corpulent—about

80 times its present mass—ignoring for a moment the possible gravitational effects on the other planets, Jupiter today might have been a red dot, scarcely brighter than a full Moon: one of those stars, indifferent and distant, to be sought out while gazing into the sunset.

CHAPTER 2

An Unexpected Promotion

The Speed of Light Becomes Absolute; Space and Time Are Bound Together Forever

It is 1905. *Annus mirabilis.* In just seven months and four papers, Albert Einstein revolutionized our understanding of the Universe and the reality to which we belong. At the turn of the century, knowledge in physics was limited and included Newton's three laws of mechanics and the principles of thermodynamics. For almost 200 years, gravity had been the only known force, but several decades earlier, in the late 1800s, it had been joined by a new one, electromagnetic force, defined by Maxwell's equations.

It was Lord Kelvin, the eminent English scientist to whom we owe major scientific discoveries such as the absolute temperature scale and the second principle of thermodynamics, who declared in September 1900 that there was nothing more to discover in physics. All that remained was to take more precise measurements.

The young physicist Albert Einstein had been working at the patent office in Bern, Switzerland, for a few years by then. It was a challenging time for Europe, but also one of profound and exciting change. New technologies were emerging thanks to electricity, transforming people's lives, especially in the cities. Photography and cinema were opening new perspectives. The interconnecting lines of the railway networks were beginning to link remote places together. In this surge of modernity, new patent applications were arriving in Bern, including those related to the synchronization of clocks.

The issue was unprecedented. Until then, if a clock in Munich and another in Bern did not show exactly the same time, it did not matter much;

the discrepancy would not have had significant consequences. However, with the spread of the telegraph, communications became much faster, and the proper functioning of railway transport required precision. It was therefore necessary that in both Munich and Bern there was agreement on when clocks should strike "midday." Solving the matter became urgent.

The correlation between space and time thus began to make itself felt for essentially pragmatic reasons. In order to synchronize two distant clocks, they had to "talk" to each other, to communicate. And the information exchanged had to travel a certain distance. Therefore, the distance between them had to be taken into account.

Einstein began to explore the possibility, and implications, of synchronizing clocks through the transmission of light signals. He devised an imaginary experiment where he introduced a quantity that had never been considered before: the speed at which observers move relative to each other.

Let us assume, Einstein proposed, that there is a Mr. M on the platform of a railway station. Two lightning bolts, A and B, fall at the same time to his right and left, at an equal distance from him. For Mr. M, the lightning strikes hit the ground at the same time. At that moment a train passes by, traveling on straight tracks at a constant speed with Mr. M' aboard, who observes what is happening on the platform from his window. The reference systems for observers M and M' chosen by Einstein for his experiment belong to a very specific category. They are *inertial reference systems,* either at rest (the platform) or in motion, but of a particular motion, rectilinear and uniform—that is, at a constant speed (the train). Because M' is moving in the direction of lightning B and away from lightning A, he will see lightning B, toward which he is heading, first; and only afterward will lightning A, behind him, reach him. Unlike Mr. M who sees the lightning bolts strike at the same instant, Mr. M' will conclude that the two lightning bolts struck at two different instants: "The observer will see the light beam emitted from B earlier than he will see that emitted by A," Einstein wrote in a popular essay explaining the theory of relativity:

> Observers who take the railway train as their reference-body must therefore come to the conclusion that lightning flash B took place earlier than lightning flash A. We thus arrive

> at the important result: Events which are simultaneous with reference to the embankment are not simultaneous with respect to the train, and *vice versa* (relativity of simultaneity). Every reference-body (co-ordinate system) has its own particular time; unless we are told the reference-body to which the statement of time refers, there is no meaning in a statement of the time of an event.

It is not the speed of the light emitted by the lightning that changes as the reference system changes (be it the stationary platform or the moving train) but rather the time it takes for that light to reach the observers, which depends on their relative motion.

The fact that the concept of *simultaneity* depends on the system of reference has enormous repercussions. If the two observers had agreed to synchronize their clocks to midday at the moment when the lightning struck the ground, it would make no difference to Mr. M which of the two lightning strikes he chose, while for Mr. M' everything would depend on it.

Galileo had already speculated about the effects that the speed of reference systems could have on the laws of physics. If a ship is traveling at a constant speed, without shaking, on a perfectly calm sea, no mechanical test carried out below deck would render it possible to determine whether the ship is moving or stationary. The laws of physics remain unchanged; they are not affected by motion. By applying the mathematical transformations that bear his name, Galileo developed the first explicit historical concept of the principle of relativity, published in 1632 in his *Dialogue Concerning the Two Chief World Systems*.

The meaning of the term *relativity* is somewhat counterintuitive. In essence, it is the search for an "absoluteness," an objective perspective that ensures interpretations of the physical world remain constant even in the face of any apparent inconsistency. Einstein was *obsessed* with identifying the mathematical transformations, the formal stratagems, that would ensure the physical laws would always work in the same way, within every reference system. This quest, this desire, lay behind the development of the two theories of relativity that heralded the beginning of modern physics.

The special theory of relativity, one of the most famous conceptual formulations of the 20th century, is explained by Einstein in a famous paper from 1905: "On the Electrodynamics of Moving Bodies" (the third of his four *annus mirabilis* papers). The link between time and the motion of observers is established by using two postulates that Einstein assumed in his imaginary experiment.

The first is the principle of relativity originally formulated by Galileo in relation to the laws of mechanics but now extended to also include the laws of electromagnetism: the laws of physics must be the same in all inertial systems, and no inertial system is privileged. The second is the principle of the invariance of the speed of light: light always travels through empty space at the same speed, in any inertial system.

Given the knowledge at the time, these two assumptions at first appeared to be irreconcilable.

Velocity was in fact deemed to be a relative quantity, varying according to the motion of observers in different reference systems. This theory, known as the composition of velocities, was also formulated by Galileo. Let me give you an example. A train passes through a station at 20 miles per hour. If I ride a scooter at 15 miles per hour in the direction of travel in the corridor of one of the carriages, a passenger on the train will confirm that I am moving at 15 miles per hour. A stationmaster looking at me from outside, through the train window, will instead say that I am moving at a speed of 35 miles per hour, which is the sum of the speed of the train and that of my scooter. If I were to ride in the opposite direction, the velocities would need to be subtracted rather than added together.

Extending this reasoning to the speed of light, *c*—from the Latin term *celeritas*—the passenger on the train and the stationmaster would see the same beam of light traveling at different speeds. Which contradicts the second assumption adopted by Einstein, that the speed of light is constant. The Galilean principle of relativity, which includes the theory of the composition of velocities, appears irreconcilable with the principle of a constant speed of light whatever the reference system. So, we find ourselves at an impasse.

Unless . . .

And here lies Einstein's great insight: unless the concept of simultaneity is sacrificed. In other words, the speed of light remains constant, and the laws of physics remain unchanged with respect to the two inertial systems that are in motion relative to each other—as the two postulates demand—if one accepts that observers M and M' do not, and never will, agree on time. In other words, two events that happen at the same instant for a stationary observer necessarily happen at different times for an observer in inertial motion.

* * *

The measurement of speeds, and in particular that of light rays, has always fascinated inquisitive minds. Aristotle recounts that Empedocles was convinced that the light emitted by the Sun took time to reach Earth, whereas Aristotle disagreed. He was more an advocate of instantaneous travel and wrote: "Light is due to the presence of something but is not a motion." Aristotle (unlike Empedocles and those like him who considered light to be corporeal and emanating from something) thought of light as a property, a state. Something that does not quite attain the ontological status of substance but is instead a mere accident or rather the happenstance of a transparent body. Light and darkness are but two different qualities of the same body, and the transition from one to the other involves no time lag; it is immediate.

Galileo was the first to investigate the question of the speed of light propagation and experimented with two lanterns, covered by a dark cloth, placed on two small hills about a mile apart. Galileo then uncovered his lantern, and his student was instructed to uncover his own lantern as soon as he saw Galileo's light up on the other hillock. Some time elapsed between the two events, but it was too short for Galileo to perceive. His intuition, however, was definitely right.

Half a century after Galileo's experiment, the first demonstration that the propagation of light is not instantaneous, and that a ray of light therefore travels from one point to another, was provided by a Danish astronomer. In 1676, while studying the motion of one of Jupiter's moons, Io, from the Paris Observatory, Ole Rømer noticed variations in the time interval between its eclipses that seemed to be due to changes in the distance between Jupiter and Earth. This effect could only be explained if light had a finite speed.

However, precise numbers were still lacking. A few years later, in his 1704 treatise *Opticks*, Newton calculated the time taken for light to travel from the Sun to Earth and found a value of between 7 and 8 minutes. Remarkably accurate for that time, considering that the actual value is about 8 minutes and 20 seconds.

Later, other scientists, such as James Bradley and Hippolyte Fizeau, refined the measurements and confirmed that the propagation of light could not be instantaneous. Light travels at a speed that is not infinite, but has a definite value of approximately 300,000 kilometers, or 186,000 miles, per second: in 1 hour, light travels just over 1 billion kilometers—more than 600 million miles. Yet, however fast, it still needs time to cover great distances.

Galileo's lantern experiment would have had better results if one of the lanterns had been on the Moon. Then the student would have been able to measure a delay in the reception of the light signal of about 1.3 seconds, which is the time it takes for an electromagnetic wave to travel the approximately 238,000 miles that separate us from the Moon, our only satellite. A delay of only a few seconds, but one that is destined to increase. The distance between us and the Moon, in fact, grows every year by about 4 centimeters. The rise of Earth's oceanic water masses due to tidal forces creates a kind of gravitational pull on the Moon. Its orbit is stretched, and it moves away from us.

The speed of light, *c*, makes its appearance in the equations formulated by James Clerk Maxwell in 1865 to describe the propagation of electromagnetic waves. In the Newtonian framework, the electric field and the magnetic field were quite distinct, and their activity involved a vague concept of forces at a distance. With his equations, Maxwell revealed two important things. First, a new perspective in which the electric field and the magnetic field are actually two aspects of the same force. Second, and surprisingly, that electromagnetic waves travel at the same speed as light. Thus suggesting—most importantly—that light is a form of electromagnetic radiation, a theoretical conclusion later confirmed by Heinrich Hertz's experiments.

Maxwell's laws defining electromagnetic force, unheard of in Galileo's time, include the speed of light in a vacuum. If, as it appeared from the measurements, this velocity is constant, then the Galilean principle of relativity, firmly anchored in the idea of the composition of relative velocities, was in danger of being overturned. Maxwell himself thought that electromagnetic waves were transmitted through a luminiferous ether and that his equations were therefore only valid in those inertial reference systems that were "stationary" with respect to the ether. There is a contradiction in there somewhere.

Shortly afterward, in 1887, the belief that light propagated via a kind of ether and that its speed depended on the motion of the observers was refuted on an experimental basis by Albert Michelson and Edward Morley. The concept behind their experiment was this: if light from the Sun and other stars propagated via an ether, the latter would necessarily permeate the entire Solar System. Thus, in its orbit around the Sun, Earth would move through the ether, and the light rays on our planet would be subject to the effects of an "ether wind." A ray of light that traveled on the ether wind, in the same direction as Earth's movement, would travel a distance that was different from that of a second ray traveling in the opposite direction. Using sophisticated equipment, the two scientists demonstrated that light, contrary to their expectations, always travels at the same speed, regardless of the direction in which it travels relative to the motion of Earth. Ether thus exited the scene, and the Galilean law of the composition of velocities proved to be no longer valid. In the new world that emerged, the speed of light is a constant, independent of the motion of the observers.

* * *

Einstein had been fascinated by electromagnetism since his early teens. When he was about 16 years old, he tried to imagine, in one of his usual thought experiments, running alongside a ray of light. As he later wrote in his autobiographical notes: "I should observe such a beam of light as an electromagnetic field at rest." In other words, he would observe a beam of light "frozen" in space. But this was not possible. Maxwell's equations define light as the motion and oscillation of electromagnetic fields. A beam of light that does not move is no longer light. This inconsistency between Galileo's relativity and Maxwell's equations was one of Einstein's obsessions, triggering in him a genuine "psychic tension" that, he later recalled, actually made his palms sweat.

His aim was to find a way of illustrating physical phenomena, whether it be through the laws of mechanics or electromagnetism (the two forces known at the time), independent of the chosen reference system.

By this time it was clear that Galileo's transformations were no longer valid. Therefore, Einstein proposed replacing them with new ones, initially put forward by the Dutch physicist Hendrik Lorentz. Starting from the invariance of the speed of light, the Lorentz transformations permitted the "magic" of keeping both the equations of mechanics and those of electromagnetism invariant in any frame of reference. In this way, it was possible to formulate a principle of relativity that could apply to all the laws of physics. A great achievement: the much desired "invariance."

This realization turned the physical world as we knew it upside down. The speed of light is a constant that does not change when moving from one frame of reference to another.

The impact of the promotion of the speed of light to an absolute constant, no longer relative to the observer but invariant, meant that it was no longer possible for space and time to be absolute or represent separate realities. The constancy of the speed of light binds them together in the unexpected fate of an eternal tango. Thus the concept of space-time came into play, and time, joining the spatial triad, became the fourth dimension.

As we know, speed (more precisely, velocity; v) is calculated by dividing the distance (s) traveled by the time (t) taken to travel it, that is: $v = s/t$. Newton's vision of absolute time and space was only compatible with a speed of light that assumes different values for observers in motion. Once the speed of light is fixed and independent of the motion of observers, space and time must be able to continuously "adjust" to each other so that their ratio remains constant and always equates to the speed of light for any observer.

With the conceptual introduction of special relativity, a new reality emerges that our five senses could not have perceived.

The famous example, again fictitious, proposed by Einstein to illustrate the relationship between the speed of light and time dilation is that of the twins. Simplifying it slightly, let us suppose that the twins are called Massimo and Stefano and it is their 18th birthday. Massimo decides to set off in a spaceship on a year-long journey to deep space and back, traveling at close to the speed of light, say at 99 percent of c. Time passes and Massimo, who has promised his twin brother to return after 1 year, looks at his calendar and, 6 months later, reverses the direction of the spaceship and heads home in time to celebrate a new birthday together. This is where things start to get weird. Looking through a telescope, Stefano observes his brother returning. And he is stunned. The spaceship, which on Earth was 100 yards long—just shy of an American football field—now appears much shorter to him, only about 14 yards long. And there are more surprises. When Massimo returns home, he expects to find 19 candles on the cake. Stefano is instead celebrating his 26th birthday. Unlike his brother, who left and then, by reversing the direction of the spacecraft, returned to Earth after a year's journey at close to the speed of light, Stefano has aged 7 years.

For two observers who are both moving relative to each other in different inertial reference systems, the time elapsed between two events is no longer the same. They will not even agree on the size of an object. Time dilation and length contraction go hand in hand. Where one occurs, so does the other. And it is not the result of a change in perspective or divergent perceptions. Reality simply alters, as is now confirmed by extensive experimental evidence.

An ideal laboratory for testing the special theory of relativity is the spectacular world of cosmic rays. These astroparticles, electrically charged and consisting mostly of protons, arrive from deep space traveling at close to the speed of light.

When they enter the atmosphere, at an altitude of about 12 to 30 miles above Earth, they collide with the oxygen and nitrogen nuclei present at those heights and burst into a cascade, a fantastic spray, of elementary particles. Among them are muons. Muons are highly unstable, with a very short lifespan of about 2.2 microseconds. After that, they decay, turning into something else. Imagine muons as blue particles that after about 2.2 microseconds, *puff!*, turn red. Given their tiny lifespan and running at close to the speed of light, they should be able to travel only a short distance, less than half a mile (about 660 meters), into the very upper layers of the atmosphere before they "die" and change color, unnoticed. One would therefore expect to see at sea level, miles below, only red particles. And yet, no. Strangely, blue particles, the muons, abound. And this is where special relativity comes in. The lifespan of about 2.2 microseconds is that measured in the laboratory, when the muon is at rest. If the muon were wearing a watch, the calculated time in its reference system before decaying would be 2.2 microseconds. For those of us observing it from another reference system, namely from Earth, the muon is not stationary but is traveling at almost the speed of light. The decay time measured by our watch would be dilated compared with that marked by the muon's own. In other words, a muon in motion lives longer than a stationary muon. Traveling close to the speed of light, its lifespan increases; it can become dozens of times longer. The particle thus has plenty of time to travel, and remain a muon, all the way down to sea level.

The dilation of the decay time, a consequence of a change in the reference system, offered a coherent and brilliant explanation for the observation of muons at sea level. Special relativity emerged from thought experiments and gave science an unexpected perspective.

* * *

The speed of light, invariant for observers in inertial systems, *can* change, however, when propagating in a medium other than a vacuum.

More precisely, the speed of light, about 300,000 km/s (186,000 mi/s) in a vacuum, decreases when it passes through other elements, such as glass or water. Being composed of overlapping electromagnetic waves, light in fact interacts with the materials it passes through; its photons are constantly absorbed and reemitted by the atoms, and this phenomenon takes time. In water, for example, light slows down and moves at about 225,000 km/s (140,000 mi/s), 75 percent of its speed in a vacuum. This changes if instead of photons we consider neutrinos, which move at a speed close to that of light in a vacuum and, interacting only very weakly with matter, are not slowed down by it. In a swimming race between a ray of light and a neutrino, the latter would therefore win.

On the matter of light's ability to *dive into* and *swim through* water, there is a little story worth telling.

"Who is that waving, do you know her?" said Niní from the other boat. "I didn't even turn around," recalls Massimo in Raffaele La Capria's *The Mortal Wound*. "I was looking at the sea floor. The right spot to dive in. What a sea there was that day!" A useful story when, like Massimo, you are looking for the right spot to dive for a sea bream glimpsed in the waters.

Because of a physical phenomenon called *refraction,* when an electromagnetic wave—light—passes at a certain angle through the surface separating two different elements, in this case air and water, the angle at which it travels changes. It becomes narrower. That is why, when we look at a teaspoon dipped in half a glass of water, it appears "bent." A ray of light that plunges from point A into the sea at a given angle will deflect its trajectory as soon as it enters the water. The angle, relative to the vertical, decreases.

Let us now imagine that, underwater, the diverted beam of light passes through point B. Its path, starting from A and arriving at B, will appear "broken" into two segments that have different angles. There is something special about the trajectory of the light beam connecting A and B. It is not the shortest distance between these two points, but it is the quickest path from A to B. If, from the beach (point A), you saw a person in the water needing help (point B), to reach him as quickly as possible you would have to dive in following a "broken" path, similar to the one light would take. It is as if the trajectory of that ray of light accounts for the fact that in water, you swim more slowly.

In other words, light has no time to waste. This is, in a sense, the exciting information contained in that change of angle.

* * *

In 1905, the true *annus mirabilis* of physics, Einstein published three other papers of fundamental importance. Before presenting his theory of relativity, Einstein had first published "On a Heuristic Point of View Concerning the Production and Transformation of Light." In it, he laid the foundation for the birth of quantum mechanics, explaining the photoelectric effect—that is, the absorption and emission of electromagnetic radiation (such as light) by bodies. Einstein proposed the revolutionary idea that light radiation, in its most intrinsic form, was composed of "light quanta." Small, bright packages. This idea, which originated from reading Max Planck's research, was at odds with Maxwell's equations. For Maxwell, in fact, electromagnetic radiation was wave-like in nature.

In truth, both scientists were right. Today, the

particle nature of light is an established fact: the "quanta of light" we call "photons." And, with that paper, Einstein laid the first brick of his studies on the photoelectric effect that earned him the Nobel Prize in Physics in 1921.

The second of the 1905 papers has nothing to do with relativity, either. It is titled "On the Movement of Small Particles Suspended in Stationary Liquids Required by the Molecular-Kinetic Theory of Heat." Here Einstein provided a theoretical explanation of Brownian motion, the phenomenon discovered by Robert Brown in 1827 regarding the motion of small particles in liquids. Brown realized that if you look through a microscope at the motion of grains of pollen in water, they appear to move in a chaotic manner, constantly changing direction. Einstein worked out why.

The pollen grains are so small that each one individually feels the impacts from the molecules that make up water. With each collision with a water molecule, the grain of pollen changes direction. It was the first time that a quantitative and mathematical explanation had been provided for this apparently chaotic phenomenon. And, with this paper, Einstein made a decisive contribution to our understanding of the microscopic nature of liquids and how molecules move within them.

Einstein's third paper of 1905 was the previously discussed "On the Electrodynamics of Moving Bodies." And lastly, the fourth paper Einstein published in 1905 was another on the theory of relativity. He posed a question already in the title: "Does the Inertia of a Body Depend upon Its Energy Content?" In this paper, he explored the consequences of the relationship between mass and energy. To understand how important this paper is, suffice it to say that in it, the most concise and renowned formula in history appeared for the first time, now famously abbreviated to $E = mc^2$.

* * *

Einstein's genius lay in his unique ability to develop a comprehensive vision of the cutting-edge theories and experiments in the physics of his time, combining them in the search for a universal interpretation. It takes curiosity, determination, and extraordinary patience to ask new questions and not settle for the first answers. Einstein was thus able to elaborate a theory that took into account the entire knowledge of physical phenomena then available, whether experimental or theoretical. It was not at all easy for a scientist of that era to navigate through so many results that seemed surprising, sometimes irreconcilable, and that seemed to preclude overall coherence. Einstein succeeded by changing his perspective, introducing new insights, and relying on his intuition. Then, through mathematics, everything changed.

In the bubble of reality we inhabit, we would never have become aware of these seemingly unbelievable relativistic effects. Time dilation and the contraction of distances begin to be faintly perceptible only at speeds greater than one-tenth the speed of light, about 30,000 kilometers (18,600 miles) per second.

It is possible to see the concrete effects of time dilation as speed changes in an aircraft. In an experiment in 1971, a high-precision watch was placed on a plane flying east, another was placed on a plane flying west, and a third was left stationary on the ground at the departure point. When the two planes returned, their respective watches showed a difference of 100 nanoseconds: a minuscule, but measurable, slowing of time.

The 1905 special theory of relativity explains only part of the time dilation observed in this experiment. The gravitational field, which is weaker at high altitudes, also plays a role. However, this effect would only be formulated by Einstein later, in the general theory of relativity.

* * *

At the heart of the revolutionary concept of space-time is, therefore, a mathematical formula. The speed of light as an absolute physical quantity binds space and time together, linking them in an inexorable and necessary relationship.

A four-dimensional space-time was first formulated by Hermann Minkowski, who announced at a conference in 1908, three years after the publication of the special theory of relativity: "Henceforth space by itself, and time by itself, are doomed to fade away into mere shadows, and only a kind of union of the two will preserve an independent reality."

In special relativity, the values of time and space differ for different observers. However, Minkowski found a way to reconcile them through a formalization of space-time that provided a mathematical explanation of events upon which all observers could agree.

In the four-dimensional representation of space-time imagined by Minkowski, the past and future are contained within two symmetrical cones, whose vertices meet at a point, the present. All physical reality is only possible within these cones, on the perimeter of which is the realm of all that moves at the speed of light. Because nothing can move faster than light, the space-time outside the cones is inaccessible. The trajectories of all physical objects are therefore contained within them.

Einstein originally thought Minkowski's idea was too complicated. However, he soon changed his mind, and the structure of space-time described by Minkowski proved to be essential for developing the general theory of relativity.

Time thus became a new dimension in addition to the three spatial coordinates. To be identified, every point in the Universe needs those three spatial coordinates and the corresponding time.

In the two-dimensional Euclidean space with spatial coordinates (x, y), the length of a ruler

L is defined by the familiar Pythagorean theorem: $L^2 = x^2 + y^2$. Now let us imagine that the two-dimensional example is represented by the ruler placed on a table. We lift one end, leaving the other resting on the table. The ruler now occupies a three-dimensional space, and to describe its length, we must add the height coordinate, let us call it z', relative to the table. The length of the ruler can be calculated by using the three-dimensional version of the Pythagorean theorem: $L^2 = x'^2 + y'^2 + z'^2$, where x' and y' are the projections of the ruler on the table.

It is difficult to imagine a length or, rather, an interval in a four-dimensional (4-D) space-time. In Minkowski's four-dimensional metric of space-time, the three spatial coordinates are always x, y, z, while time t is "spatialized" by multiplying it by the speed of light to give it the same dimension of length shared by the three spatial coordinates, allowing them to be consistently added together or subtracted from each other. In Minkowski's space, each point is therefore identified by the tetrad (x, y, z, ct).

The Pythagorean formula can thus also be extended to encompass a four-dimensional space. The equation to do this, a kind of avatar of the Pythagorean theorem, allows us to calculate the space-time interval.

A small change in any quantity Q can be formalized in mathematics by the symbol dQ. The letter "d" signals the difference, or rather differential, of the quantity Q between two infinitely close points. The temporal distance between two neighboring points A and B can then be denoted by $dt = (t_B - t_A)$ and the spatial distance by $dx = (x_B - x_A)$, and the same applies to dy and dz.

Using the avatar of the Pythagorean theorem in a four-dimensional space-time, where t is substituted by ct, the distance s between a pair of neighboring points A and B is calculated, by using Minkowski's metric, as $ds^2 = c^2dt^2 - dx^2 - dy^2 - dz^2$. By introducing the three-dimensional version of the Pythagorean theorem for spatial coordinates $dl^2 = dx^2 + dy^2 + dz^2$, this is simplified to $ds^2 = c^2dt^2 - dl^2$. This short equation is a kind of non-Euclidean length calculated in space-time, referred to as the space-time interval ds.

The distinction between space and time is maintained by the very fact that the space term is preceded by a minus sign. In a four-dimensional space-time, the Pythagorean theorem takes an unusual form: the square of the hypotenuse is equal to the square of one side minus the square of the other. This minus sign, intrinsic to Minkowski's metric construction, may seem surprising, but it is the secret of the different manifestations of time and space in nature. And within the framework of special relativity, the numerical value of the space-time interval will therefore not be affected by a change of inertial system.

Returning to Einstein's famous moving train, the man on the train and the stationmaster standing on the platform will certainly have differing opinions on the time elapsed between the two events. But if they calculated the space-time interval $\mathrm{d}s$ instead of the time difference, they would both agree on the result! When it comes to the space-time interval, we can change our (inertial) reference system like a shirt. The choice of a new frame of reference obviously changes the value of the coordinates but not the interval between two points in space-time. This space-time interval is the same for all inertial observers. And this is another brick in the theory of relativity, an absolutism crystallized in the minus sign in the equation $\mathrm{d}s^2 = c^2\mathrm{d}t^2 - \mathrm{d}l^2$.

Amazing! A new invariant. Invariants are so

valuable that some claim that physics is essentially the search for them. They introduce simplicity, abolish privileges between frames of reference, and get everyone to agree.

The special theory of relativity, also called restricted relativity, published in 1905, thus describes how the Universe works at very high speeds. Space and time bind together to keep the value of the speed of light constant. Time is perceived differently by different observers and flows differently according to their relative speeds. Simultaneity no longer exists. Time can dilate and lengths can contract. It is within this context that the formula $E = mc^2$ emerges.

A new, unbreakable relationship joins the eternal bond between space and time, that between mass and energy. As Einstein himself said, "It followed from the special theory of relativity that mass and energy are both but different manifestations of the same thing—a somewhat unfamiliar conception for the average mind."

We all know from experience how difficult it is to push a full shopping cart toward the supermarket checkout. In Newtonian mechanics, the energy (E) needed to overcome the inertia of a body and move it at a certain speed v (velocity) is given by the equation $E = mv^2/2$, where m is the mass of the body. Theoretically, any velocity can be reached if an adequate amount of energy is used.

This is not so in special relativity. It is not possible to attain *any* velocity because it is not possible to go faster than light. As you approach the speed of light, it becomes ever more difficult to increase the velocity of a body. And to accelerate it to the speed of light would require an infinite amount of energy.

If more and more energy is needed to accelerate a body close to the speed of light, the question arises as to where the energy injected into the body that is not transformed into speed ends up. The answer: it becomes mass. The only other quantity involved.

Here, then, is one of the most counterintuitive consequences of special relativity. In Newtonian mechanics (and thus in everyday experience), the mass of a body remains constant, and its velocity and acceleration are entirely dependent on the force applied. In the revolution of physics brought about by Einstein, the mass of a body is no longer a fixed constant but increases as a result of the energy that is not converted into velocity.

As the speed of a body approaches that of light, its mass increases more and more. And the greater the mass, the harder it becomes to accelerate its motion.

The consequences are extraordinary. Mass and energy are no longer two separate entities but can transform into each other according to the formula $E = mc^2$. The conversion factor is the term c^2. Mass $m = \gamma m_0$ is to be understood as the mass of a body at rest, m_0, which is an invariant, multiplied by a gamma factor (γ) that increases the closer the particle's speed approaches that of light. Mass and energy prove to be two manifestations of the same reality. Once again, two concepts that were independent in Newtonian physics are now inextricably linked. Mass can be converted into energy, or rather, the energy trapped in mass can be released. In turn, energy can produce mass. And one factor does not change: the speed of light squared. If 1 kilogram (2.2 pounds) of mass were to be fully converted into energy, this would yield an amount of energy equal to that produced over three years by a gigawatt nuclear power plant, enough to power hundreds of thousands of homes.

And, as we have said, energy can be transformed into mass. This is what happens in particle accelerators and during violent astrophysical phenomena such as supernova explosions. Two photons (quanta of light that possess energy and momentum but no mass when at rest) can give rise to several pairs of particles, whose mass will depend on how much energy they possess. Particles with mass, such as an electron-positron pair or a proton-antiproton pair, can therefore appear "out of nowhere." And their properties will depend on the energy of the photons from which they emerged: $E = mc^2$. This is the magic formula that allows collisions in particle accelerators to generate new particles and antiparticles, transforming kinetic energy into mass energy.

As the famous French astrophysicist Michel Cassé would put it, a movement becomes an object, an attribute becomes a body.

Even the Sun shines thanks to the formula $E = mc^2$. Our star that illuminates our days is powered by nuclear reactions that basically turn four protons into a helium nucleus. However, four protons weigh about 0.5 percent more than a helium atom. This difference in mass becomes energy, light that is radiated away. In stars, matter is transformed, even if only partially, into light. In the primordial Universe, the inverse magic is observed: energy becomes matter. The miracle, the wonder, of *giving birth to the world,* just like in the title of a Boetti painting.

Special relativity draws a radically transformed picture of the physical world compared with that of classical physics. More complex and certainly more complete. In fact, absolute space and time no longer exist but are replaced by new absolutes, the speed of light and the space-time interval. And the foundations of the Newtonian framework for physical events disintegrate along with them: the illusion of absolute motion and absolute simultaneity. All motions are relative to a system of reference. There is no inertial system that has priority over any other one. The absolute limit of the speed of light gives rise to the equivalence of mass and energy, which leads to the exponential increase of relativistic mass when attempting to accelerate a body to the speed of light.

The basic principles of special relativity are based on systems moving toward each other in a straight line and at constant speed. However,

Einstein sought to extend his thinking on relativity to also include accelerated reference systems in his formulation. And to do so, he had to start with that question that Newton had left unanswered, entrusting it to the reader. What is the true nature of gravity?

CHAPTER 3

Gravity Changes Its Nature

And Becomes Geometry

Newton's law of universal gravitation stated, in a nutshell, that the attraction between two bodies depends on only two attributes: their mass and the distance between them. However, what was not clear was how bodies millions of miles apart were able to influence each other's motion. Newton posed this question in his *Principia,* and he left it unanswered.

More than 200 years later, the question reemerged in the context of a new perspective:

the astonishing promotion of the speed of light to an absolute value, with its new regal status as an insuperable limit. The idea that gravitational attraction between two bodies could occur instantaneously was finally ruled out.

When Einstein began to take an interest in the subject, the scientific community did not seem particularly keen on opening such a debate. Yet, despite the fact that the century had begun with Lord Kelvin's dire prediction that physics would no longer hold any surprises, Einstein's approach—mixing perspectives with imaginary experiments, overturning absolutes, transforming time, and constructing narratives of the world never told before—had already taken us far. Special relativity had established that no signal or form of "influence" could travel faster than light. A gravitational attraction acting between two distant bodies thus necessarily entails a waiting time. It was now essential to understand how this attraction, this ability to influence each other remotely, could occur. Seeking clarity on this subject was not Einstein's only goal. At the heart of his ambition lay the quest for a theory of relativity that could extend, *generalize,* the special theory of relativity to include accelerated frames of reference alongside inertial frames. Such a theory would then make the laws of physics invariant regardless of the observers' motion and the characteristics of their frames of reference.

This crucial revelation came in 1907, two years after the elaboration of special relativity, while he was at his desk at the patent office in Bern. It was a remarkable insight because it established an unexpected connection, an unimagined link, between two seemingly distant concepts. Between a force and an attribute. Between gravity and acceleration. Their equivalence was the magic formula that would redesign the Universe, even though calculating the effects would still take another decade of work.

To visualize the relationship between gravity and acceleration, let us start with an example. Let us imagine that we are inside an elevator that has no windows. Our feet are firmly on the ground and, even without being able to see outside, we can readily assume that the elevator is stationary and that our feet are where they are as a result of Earth's gravity. Yet, for the same given circumstance, there could be another explanation. We could be in the same elevator but traveling in deep, empty space at an acceleration equal to that of Earth's gravity. Due to the effect of acceleration alone, our feet would remain firmly planted on the elevator's floor. It is basically the same feeling we get when a car suddenly accelerates and we are pushed backward or when an elevator speeds up and we feel a downward pull. Let us now imagine a different scenario, with a reversed dynamic. We are in an elevator that has no windows, plunging from the 30th floor. We are therefore in free fall, and, without any handholds, we are floating in the air. But who can assure us that we are indeed in an elevator whose cable has broken? We might just as easily be in a spaceship that, in the absence of gravity, travels through space at a constant speed, and, with nothing to hold on to, we find ourselves floating in the air.

So, to recap: Being stationary with our feet firmly on the ground in the presence of gravity is equivalent to traveling through space in an accelerating spaceship. Being in free fall inside a falling elevator is akin to being in a spaceship traveling freely in space at a constant speed. Brilliant. Yet this is highly counterintuitive. Special relativity established the principle of equivalence for inertial systems, and here we find the same equivalence for accelerating systems, simply by including the force of gravity. The special relativity valid for inertial frames becomes general relativity by extending its range of application to accelerated systems as well.

The equivalence between gravity and acceleration—therein lies Einstein's great insight. As he himself wrote some time later:

> I got the happiest thought of my life in the following form: . . . the gravitational field has a relative existence only. . . . Because for an observer in free-fall from the roof of a house there is during the fall—at least in his immediate vicinity—no gravitational field. Namely, if the observer lets go of any bodies, they remain relative to him, in a state of rest or uniform motion.

In other words, in a free-falling reference system, it is not possible to perceive the presence of Earth's gravity in any way. Thus, the force of gravity does not seem to be a real force, but only an apparent one, because its effect is no longer felt when a convenient reference system is chosen.

Simply put, in a free-falling system, the force of gravity disappears, just like a magic trick.

In this game of mirrors, triggered by Einstein's great intuition about the equivalence of gravity and acceleration, reality and appearances are reconstructed from a new angle, and the pieces of the jigsaw puzzle he had been working on for so long begin to fall into place with magnificent simplicity. This left unresolved, however, the critical issue of the impossible simultaneity of gravitational attraction between bodies: certainly no small matter.

* * *

Einstein mulled over the following dilemma. There is a distance of 93 million miles between the Sun and Earth. And this space, which had been considered nothing more than a passive background, may, in fact, play a role in the dynamics of their mutual attraction. It is worth noting that physicists had, not long before, studied a similar problem, but it involved two electric charges attracting each other. They had concluded that the existence of an electromagnetic field, which permeates the space in which the charges are immersed, *informs* them that they must attract each other. Einstein started from this formulation but took it further: it is not the gravitational field that, by permeating space-time, communicates and *informs* bodies of the presence of gravitational force. Instead, space-time itself—and this is the incredible conceptual leap—*is* the gravitational field.

To advance his research, Einstein had to learn new mathematical methods, overcome deep-rooted prejudices, and study alternative geometries. In November 1915 he wrote four papers, each published a week apart, that defined the scope of the revolution of general relativity. The final version of the general theory of relativity was then published in May 1916 with the title "The Foundation of the General Theory of Relativity."

The shift in perspective was radical. Einstein argued that in the absence of matter, space-time could be thought of as a rubber membrane, smooth and taut. If we were to place a ball on the membrane and give it a forward push, the ball would move along a straight trajectory. However, if we were to place a massive marble sphere in the center of the elastic membrane, it would deform, creating a kind of depression. The ball rolled forward would then no longer go in a straight line but would instead follow this new curvature.

Let us think about the Universe in the same way. If we place the celestial bodies on top of the smooth, taut rubber membrane, it will deform. The resulting shape will determine the motion of the objects moving along it. Defining this flexible and dynamic landscape gives the force of gravity a geometric connotation, distancing it from the Newtonian idea of a mysterious action at a distance. In the context of general relativity, the motion of two bodies attracting each other is nothing more than the effect of the curvature of space-time deformed by the very presence of mass.

We can now reconsider the movements of the celestial spheres in this light. Earth revolves around the Sun not because its orbit is the result of a balance between the force of gravity and inertial velocity. Instead, its motion is the natural trajectory of a ball thrown in a straight line along an elastic surface that finds an imaginary valley in its path—space-time curved by the presence of our star. Earth's path bends, and it rolls along a groove in space-time traced around the Sun, in a perpetual orbit. The nature of gravity is revealed. No longer a force that pulls and holds a body that wants to flee, but a trajectory shaped by the curvature of space-time, a line along which objects glide effortlessly.

The Universe becomes a soft and dynamic landscape filled with valleys and dips, curved lines in which objects move fluidly, no longer battered by the pull and push of gravity. The roles change. And that king of them all, gravity, is transformed into a sinuous design woven by the hand of matter and energy into the elastic fabric of the Universe.

It took Einstein 10 years to determine exactly how space-time is warped by a given amount of matter or energy. For a long time, he worked on field equations from a geometric perspective, which enabled him to calculate the trajectories of everything moving within space-time with great precision—planets, satellites, but also rays of light whose trajectories must necessarily bend as they pass through the valley of space-time created by a star. Because yes, even a ray of light, despite being made of photons that have no mass, can "feel gravity," since its trajectory follows the buckling geometry of space-time. What a surprise that would have been for Newton. To be capable of *falling* even without having any mass. A ray of light as an apple.

Experimental proof that a ray of light bends its path when it is close to a large mass came in 1919, during a total eclipse of the Sun. The Royal Astronomical Society sent two expeditions to observe it from two very distant locations: Arthur Eddington, director of the Cambridge Observatory, left for Prince Island in the Gulf of Guinea, while Andrew Crommelin of the Greenwich Observatory went instead to Sobral in Brazil.

Solar eclipses were already regarded, during previous measurement attempts, as ideal opportunities to observe the deflection of light from distant stars. Indeed, if the position of certain stars in a night sky were to be compared with their position observed during a total eclipse, the effect of the curvature of space-time on the trajectory of light due to the presence of the Sun would be apparent. If Einstein was right, the positions of the stars during an eclipse would appear slightly different from those measured at night. On May 29 the good news reached the Royal Astronomical Society via a telegram from Eddington: the deflection had been observed. Eddington's academic authority was asserted, and those images were considered the first experimental verification of the general theory of relativity. The results, announced on November 7, 1919, swept the world, filling the front pages of newspapers and establishing Einstein's popularity. The *New York Times* reported the news with great enthusiasm: "Einstein Theory Triumphs. Stars not where they seemed or were calculated to be, but nobody need worry."

Light bends in the presence of large masses. But what about time and its passing? In the general theory of relativity, we discover that gravity also influences the flow of time, just as it was affected by motion in special relativity. Within the imprint that mass and energy leave on spacetime, time naturally bends as well. It slows down.

The closer you are to the source of gravitational attraction, for example the center of Earth, the slower a clock marks the passage of time. The result? Time passes more slowly near the sea than at the top of a mountain. And it passes more slowly on Earth than on a plane or on the ISS. The effect may be imperceptible, but it is real. The time measured by the Galileo satellite system at 14,000 miles from Earth is not the same as that of us motorists. It must therefore be corrected in order to give our GPS accurate measurements and get us home in time for dinner.

My grandfather used to say that there were two things that increased the chances of living a long life: always carry a chestnut in your pocket, and run every day on Serapo Beach in our hometown of Gaeta, Italy. Southern Italian eccentricities perhaps. But at least on the second point Einstein would have agreed with him. Moving (special relativity) and being on the beach, and therefore close to Earth's center of gravity (general relativity), are good strategies to slow down time.

General relativity explains how space and time, which are now intertwined, are changed by the presence of large masses. An interpretation of the world that has progressed very far, thanks to the kind of open-mindedness that a scientist needs to pursue and define points of no return. When you look at things from a different angle, questions take shape that were not apparent before. This was Einstein's great skill, a free, stubborn, and almost insolent curiosity. Thus, through general relativity, the impossible is transcended. Everything becomes new.

The formulation of the concept of a Universe with a flexible structure, modulated by the presence of matter and energy, settled the issue of the simultaneity of gravitational attraction. If we remove the Sun, the space-time around it, which until then had been somewhat crumpled by its presence, would stretch out again. It would become a smooth surface once more, and as it flattened, it would let Earth drift away. But none of this would happen immediately. The crumpled space-time, in fact, would take time to flatten out, while Earth, unaware, would continue to orbit the absent Sun.

The "elasticity" of space-time, the speed at which information about the Sun's disappearance would travel, would be equal to that of light. So, the effect would appear to be instantaneous, even though it would not be. If the Sun were to suddenly disappear, we would only see it disappear about 8 minutes later. And at that exact moment we would fly away.

In the general theory of relativity, the speed of light is also in a way the speed or "slowness" of the reactions of space-time. That magic number is therefore not only the absolute and insurmountable attribute of the motion of an electromagnetic wave; it also defines the Universe itself, the movement with which its fabric stretches or deforms, the elasticity with which its fibers react, and how information propagates within it.

The geometry of the Universe, its inherent structure, is a dynamic entity that deforms, ripples, twists, and, in some places, sinks. It all depends on the energy and matter that fills it. To quote a famous phrase attributed to the American physicist John Wheeler, "Spacetime tells matter how to move; matter tells spacetime how to curve."

* * *

Right up until the first direct measurement of gravitational waves in 2015, much of what we knew about the Universe, at least from a certain point onward in its evolution, was taught to us by light or its absence. But gravity often intervened to complete the story, revealing what tries to hide itself, flushing out unseen presences. Unearthing black holes.

A distinctive feature of a black hole in the soft tissue of space-time is the event horizon. It is a terrifying threshold, a line of no return that marks the entrance to the vortex of space-time created by the presence of an enormous amount of mass concentrated in a very small space. Beyond this limit neither matter nor radiation can escape.

In a black hole, the immense gravitational force, which is billions of times that of Earth, traps information and light, and the extreme deformation of the space-time curvature causes an extraordinary slowing of the flow of time. If we were to spend 1 year in orbit at 100 meters from the event horizon of the black hole Sagittarius A*, which lies at the heart of our galaxy and has a mass 4 million times that of the Sun, 11,000 years would pass on Earth during the year of our orbit.

The possibility of a Universe populated by black holes is part of the revolution sparked by Einstein's equations. Once again, however, the emergence of a new perspective was met with skepticism. Einstein himself was disinclined to believe that black holes could be real objects and needed more time to come to terms with what was flowing from his pen. Just like with the expansion of the Universe.

A few weeks after the November 25, 1915, publication of the last of the papers that laid the foundations of general relativity, "The Field Equations of Gravitation," another step forward was already being taken in a remote part of the world that was in turmoil: the frontline trenches of the First World War. The German scientist

Karl Schwarzschild, a prominent astronomer and director of the Potsdam Astrophysical Observatory, decided to enlist at the outbreak of the war even though he was more than 40 years old. And so he found himself fighting on the Russian front as an artillery lieutenant.

While in the trenches, he received a package from his friend Einstein containing "The Field Equations of Gravitation." Despite the war and the agony of a rare skin disease he had recently contracted, Karl Schwarzschild sent Albert Einstein a letter a few days later. It was December 22, 1915. In his letter, the scientist gave Einstein the exact solution to his field equations of gravitation, which became known as the Schwarzschild metric. This solution provided a way of describing how space-time bends around spherically symmetrical objects, such as planets, stars, but also black holes, whose extreme density can generate gravitational fields so intense that they even trap light. The letter concluded with these words:

> As you see, the war treated me kindly enough, in spite of the heavy gunfire, to allow me to get away from it all and take this walk in the land of your ideas.

Einstein was astonished; he himself had found only approximate solutions and, as he then wrote, did not expect that the exact solution to the problem could be formulated so simply. Schwarzschild died only a year later, at the age of 42, most likely as a result of his illness.

Thus, black holes materialized through the language of mathematics. A solution that through its simplicity delivered the invisible and confirmed its beauty.

Initially, these bizarre objects were called dark or "frozen" stars; the term *black hole* was only later popularized by Wheeler in 1967. The event horizon played a fundamental part in Schwarzschild's solution, yet for a long time, black holes were considered essentially an elegant mathematical oddity. The actual existence of a black hole requires extreme conditions. For example, for a star such as the Sun to become a black hole, its mass would need to be compressed to an area no larger than the city center of Rome. If, instead, its mass were comparable to that of Earth, it would need to shrink to the size of a blueberry.

It is not easy to imagine a physical mechanism capable of triggering the incredible compression of a star into a very small space.

Until then, it was thought that the fate of every star was to end its days as a small, dense, white dwarf. The scientist Subrahmanyan Chandrasekhar was only 19 years old when in 1930, during a long sea voyage from India to England, he formulated the theory of a more complex dynamic. Taking into account recent developments in quantum mechanics that helped explain the behavior of gases made up of electrons and protons, the young scientist calculated that when a star has a mass at least 1.4 times that of our Sun—a number now known as the Chandrasekhar limit—it can end its life cycle in various ways: by exploding into a supernova, by exploding and then collapsing into a neutron star, or by turning into a black hole. But he was not taken seriously. Commenting on Chandrasekhar's hypothesis at the meeting of the Royal Astronomical Society in 1935, Arthur Eddington asserted: "There should be a law of nature to prevent a star from behaving in this absurd way!"

Chandrasekhar calculated that when stars that are much larger than the Sun, 10 or 20 times heavier, are no longer supported by the heat of combustion, they collapse, crushed under their own weight until they curve space so dramatically that they disappear into deep chasms. So what was once a big, beautiful star is transformed into a stellar-mass black hole. An abyss of space-time created by the swallowing up and compression of stellar debris into a very small space.

It seemed an incredibly complex mechanism. The idea that black holes were real objects was for many years considered irrational. And Einstein was among the many skeptical scientists. The concept was too radical, "unconvincing," as he wrote in an article published in the journal *Annals of Mathematics*. Hidden spots into which the Universe sinks and is snuffed out. In another part of the world, in 1939, Eugenio Montale conveyed that sense of faltering on the edge between light and darkness in just a few words: "The life that gleams / is the only one you see."

Today we know that the very fabric of the cosmos is dotted with these precipices, of varying size and origin, within which the Universe hides. The effect of gravity is so intense that nothing, not even light, can climb up out of these abysses and escape. The smallest black holes are thought to be the size of an atom with the mass of an asteroid, objects that have probably been present in the structure of space-time since the beginning of the Universe. Then there are stellar-mass black holes, which are formed at the price of the death of a star, each having a mass tens of times that of the Sun.

Finally, the so-called supermassive black holes are enormous, weighing up to tens of billions of solar masses. Possibly created along with their host galaxy, they are voracious and tend to "swallow" nearby stars, growing progressively larger. Today, it is believed that virtually every galaxy harbors a black hole of millions or billions of solar masses.

Although the term *black* implies darkness, black holes can reveal themselves through sharp edges that define the boundary of that darkness, marked by sudden changes in light, around which matter flares up and glows.

In April 2019, the first image of a black hole—or rather, of the light emitted by material orbiting the black hole at the center of the supergiant galaxy Virgo A—reached us. M87* is an extraordinary object, as massive as 6.5 billion Suns, and it sits 55 million light years from Earth. It was not captured with a real photographic *click*. It took 2 years of the international Event Horizon Telescope (EHT) project and the close collaboration of more than 200 researchers worldwide to achieve this incredible result.

A dark shadow, the unseen abyss of a black hole, where light is captured forever, surrounded by a spectacular orange-red radiation emitted by matter before it crosses the so-called point of no return. The image appears distorted because the extreme gravitational force of the black hole warps the surrounding space-time, bending the path of light and creating a magnifying-lens effect that makes the shadow appear larger than it really is.

We did not have to wait long for a second, astonishing "photograph": that of our own "personal" cosmic monster at the heart of the Milky Way, Sagittarius A*, some 27,000 light years from our Solar System. May 2022—what a sight! The first direct proof of its existence: a magnificent new image obtained by the EHT that captured the black hole sitting at the very center of the Milky Way. A mass approximately 4 million times that of the Sun concentrated in an area that has a radius of about one-tenth the distance between the Sun and Earth. Despite the fact that these two black holes, the first to be photographed, are located in galaxies distant from each other and have very different masses—M87* is about 1,600 times heavier than Sagittarius A*—the laws of physics established by general relativity around these mysterious objects seem to work the same way. Andrea Ghez, who in 2020 was the fourth woman to win a Nobel Prize in Physics, had already confirmed with her research that photons emitted by stars orbiting Sagittarius A* behave in exactly the way Einstein described.

These thrilling images from the EHT once again pay homage to that century-old intuition. An orange glow around a dark heart. Blurred and distant burning crowns.

* * *

Space-time bends. It deforms. But not only that. Incredibly, it vibrates. It can quiver like the surface of water swept by a breeze. Einstein reached this conclusion between 1916 and 1918. Just as a moving electric charge produces electromagnetic waves, large amounts of moving matter or energy can also produce gravitational waves. The explosion of a supernova can release tremendous energy that, in the form of vibrations, ripples the surface of space-time. Note, however, that gravitational waves do not travel through space the way light rays or sea or sound waves do. A gravitational wave is a puff rippling through space-time, a curving tremor traveling at the speed of light.

Capturing a gravitational wave had long been one of the missing pieces in confirming the general theory of relativity. The technical challenges were immense, considering that the oscillations of a gravitational wave are infinitesimal.

But thanks to the joint work of the Laser Interferometer Gravitational-Wave Observatory (LIGO) in the United States and of Virgo in Italy, near Pisa, the long-awaited confirmation finally arrived on the morning of September 14, 2015. For the first time, the motion of a gravitational wave had been measured. An imperceptible movement, the amplitude a thousandth the diameter of a proton. An imperceptible movement, a gigantic discovery.

The Universe revealed itself this time with a shiver. While we can "hear" thanks to sound waves and "observe" thanks to electromagnetic waves, gravitational waves add a powerful and hitherto unknown perception of the world. A new "sense." And with it, the opportunity to gather information that had remained inaccessible since the beginning of time.

The source of that tiny ripple was a remote event that occurred about 1.3 billion years earlier. Two enormous black holes, one of 29 solar masses and the other of 36 solar masses, collided. They had danced around each other for a while, then ended their journey with a spectacular collision.

The fusion of the two black holes gave birth to a monster, a black hole of 62 solar masses. And the 3 solar masses missing from the tally became energy released into the fabric of space-time in just 200 milliseconds—a power 50 times that emitted in the form of electromagnetic waves in that same time frame by all the stars of all the galaxies in the observable Universe. The most powerful explosion ever observed, second only to the Big Bang. A wake of gravitational waves thus began spreading through the cosmos, gradually weakening as it did so. The LIGO scientists, having received and processed the signal, wanted to *listen* to it. Or rather, turn those oscillatory vibrations into the sound that identifies the violent and extraordinary collision between two gigantic black holes. Surprise. Something unexpected emerged. A magnificent chirp. The term, as reported by Stefan Helmreich in his article "The Cosmic Chirp," originated with radar research engineers in the 1950s, who compared a signal that had a sudden rise/fall in frequency to the chirping of a bird. The expression was later used in 1951 in a Bell Lab memo titled "Not with a Bang, but a Chirp," paraphrasing the final line of a poem by T. S. Eliot, "The Hollow Men": "This is the way the world ends / Not with a bang but with a whimper."

The "capture" of the gravitational wave had a further, very important, scientific relevance. It was the first direct proof that black holes actually existed.

Today there are three instruments aimed at detecting gravitational waves. In addition to LIGO, there is also the aforementioned Virgo and the KAGRA (Kamioka Gravitational Wave Detector) in Japan. But the prospect of two orbiting interferometers is on the horizon. The installation of the first one, DECIGO (Deci-hertz Interferometer Gravitational-Wave Observatory), is planned for 2027, with LISA (Laser Interferometer Space Antenna), a collaboration between ESA and NASA, due to join it in 2037.

Compared with those built on Earth, orbiting interferometers can have much longer arms, allowing us to probe longer wavelengths and observe phenomena occurring at lower energy levels. The LISA project, for example, is planning arms of 1.5 million miles, about seven times the average distance between Earth and the Moon!

Like gravity, gravitational waves are present in every corner of the Universe, and nothing can stand in their way. The Universe lets them pass, unlike light, which, as we have seen, can be absorbed by the obstacles it encounters.

While light gives us information about the source from which it comes, gravitational waves can tell us even more. Slipping along almost undisturbed, they stretch and crush the objects they find in their path, keeping a memory of these passages. They can then tell us about what they encountered along their journey. The information that gravitational waves carry, together with the information we receive through light, opens up entirely new perspectives for understanding our Universe. A new field is born: multi-messenger astronomy.

No matter how powerful a telescope may be, there is nevertheless a kind of opaque screen located some 380,000 years after the Big Bang that hides wonders and mysteries of earlier times. In fact, light only emerged at that point.

However, the Big Bang may have released a trail of oscillations, and this echo made of gravitational waves may still be traveling from the beginning of time. If we could listen to it, we would understand more about that long and beautiful chapter in the history of the Universe that began well before the emergence of photons.

Since 2015, the LIGO and Virgo laser interferometers have picked up several dozen signals produced by the fusion of black holes and neutron stars. Future experiments with third-generation interferometers, such as the Einstein Telescope and the Cosmic Explorer, but also that of the LISA space mission, will be so sophisticated that they will certainly offer new perspectives. A growing number of windows from which to observe and follow the journey of those tiny, hypnotic chirps, which can tell us additional, even unexpected, stories.

* * *

The observation of the collision of neutron stars has also clarified mysteries surrounding the formation of heavy elements. The snail on the windowsill, grandmother's desk, that flower in the garden, the bee fluttering on it, a margherita pizza, butterflies, a rock in the sea. We are all made of the same stuff: atoms. Produced in the first moments after the Big Bang, or in the heart of a star, or in one of those wonderful explosions that spell its end. About 10 percent of our body is made up of primordial elements: we have within us a memory of the beginning of time.

Most of the other elements, on the other hand, originated from the stars. Nuclear fusion within stars is not only responsible for their marvelous brilliance but is also a forge in which a good portion of the elements of the periodic table are produced.

In that fiery heart, hydrogen atoms join with helium atoms, which in turn fuse with other atoms and produce new, heavier elements of the periodic table, such as carbon, oxygen, calcium, and many others, so important for human life. Right down to iron. And then something happens. The energy produced within the stars becomes insufficient to create heavier elements.

Yet there are many chemical elements heavier than iron in the periodic table. The zinc we find in tomatoes, the copper in the pots and pans in grandmother's kitchen, the gold and platinum that sparkle in jewelers' windows. These elements do not come from the inner core of stars but from the fate that awaits them when, having exhausted their nuclear fuel, they die. Once the thermonuclear reactions inside a star cease, gravity prevails and matter collapses. Their transformation into new celestial objects will depend on their mass.

Our Sun, for example, like other stars of similar mass, will die a quiet death. When it has used up all its hydrogen, in about 4.5 billion years, it will undergo a protracted agony. It will become a red giant and will expand enough to swallow Mercury and Venus and probably our own planet as well. Then, after another 1 or 2 billion years, it will collapse to become a white dwarf, small (about the size of Earth) and extremely dense (a teaspoon of white dwarf has a mass of several tons), until it becomes almost invisible.

Massive stars (about eight times the mass of the Sun) die faster and more dramatically. After several transformations, the central nucleus, increasingly hot and dense, explodes, and they end their lives as magnificent supernovas. It is the most beautiful and violent explosion known to the Universe. A magnificent spectacle—extraordinary fireworks—shooting the materials produced within the star in every direction and above all with enormous energy. And so new collisions between the nuclei begin to occur, leading to new combinations that end up producing the heaviest elements after iron. Which is how the periodic table became enriched with new elements.

As we have seen, stars significantly larger than our Sun, with masses that are 10 or 20 times that of the Sun, succumb to their own gravity when they are no longer supported by the heat of fusion. This collapse warps space-time so profoundly that the matter sinks into it. And thus the magnificent black holes that dot the Universe emerge.

Yet there is one type of object in the sky that plays an especially important role in the production of new atoms. These are the neutron stars, which are extremely dense and compact clusters. Imagine a mass slightly greater than that of the Sun compressed into a sphere with a radius of about 6 to 12 miles. In 2017, the effect of the collision between two neutron stars could be measured for the first time thanks to the observation of gravitational waves. A great many heavy elements were seen to escape from this collision, which released incredible amounts of energy. And, surprisingly, platinum and gold also appeared in huge quantities, the equivalent of dozens of times the mass of Earth. Here, then, is the mechanism from which all the gold in the world is formed, and the platinum of the queen's tiara. A wonderful, distant, and unexpected cosmic collision.

"Be plural like the Universe," Fernando Pessoa exhorted. The Universe is undoubtedly a plural, turbulent world that fuses and recomposes, full of assaults and storms. Alchemies and mechanisms from which—and here is the beauty of it—other wonders emerge.

Galaxies, for example, are not isolated islands floating unaware and lonely in a dull, eternal night. Gravity likes to play with them. So they embrace, merge, or tease each other by brushing up against each other and stealing a few stars. But that is not all. The larger ones may even swallow the smaller ones. It is believed that there are few galaxies in the Universe that have not suffered brutal events of this kind.

Only if two galaxies hurtle past each other at a speed greater than that at which they are escaping from their respective gravitational fields—which is the speed at which they would become inextricably linked and begin their dance toward fusion—does it end there, probably without consequences. The stars, planets, and worlds housed in each of them might then remain oblivious.

In this tumultuous Universe, even our beautiful galaxy has had a turbulent life and has grown over time, attracting smaller galaxies and clusters of stars, which have since merged with the rest. The data from ESA's Gaia mission demonstrate this clearly. The Milky Way is currently, and very slowly, swallowing up the dwarf Sagittarius Galaxy and its few tens of millions of stars. And the stellar content of the Milky Way's inner ring appears to be dominated by the debris of dwarf satellite galaxies, such as Gaia-Enceladus, which merged with the Milky Way some 8 to 11 billion years ago. The attraction between the Milky Way and Sagittarius has been going on for a long time, and as various international studies based on the Gaia data show, these encounters have also played an important role in the evolution of our galaxy. The first close fly-by was around 5 to 6 billion years ago, followed by another about 2 billion years ago, and then a third around a billion years ago. So for billions of years now, Sagittarius has been visiting us every now and then, bringing us new stars and amusing itself by altering our form and shape a little each time. And it looks like one of those forays did indeed affect us. The Sun, which formed some 4.6 billion years ago after the collapse of a huge cloud of gas and dust, may be one of the stars born during that first gravitational interaction with our little neighbor.

Galaxies move away and come closer. As we will learn, Edwin Hubble deduced the expansion of the Universe by observing the redshift in the galaxies' light, indicating they are gradually receding.

Yet, if we look at the Andromeda Galaxy, another neighbor, we see that its light is not redshifted, but blueshifted. *Achtung!* Andromeda is approaching. Andromeda and the Milky Way are the largest of the several dozen galaxies that make up the Local Group. Andromeda is more massive, boasting trillions of stars, while our galaxy is more modest, with just a few hundred billion stars. They are quite close, even though the expansion of the Universe is driving them apart at about 37 miles per second. But that is not enough to halt the inevitable fatal attraction.

Andromeda, which is 2.5 million light years away from us, is heading straight for us. The Milky Way and Andromeda are in fact rushing toward each other at about 70 miles per second. In some 4.5 billion years, the encounter—the embrace? the collision?—between these two galaxies, the first to populate our sky, will be spectacular. And silent. In a boundless, empty space that knows no sound. A slow, poetic movement, which, without so much as a *Bang!*, will transform existing worlds into something else. It is a deterministic and written future from which we cannot escape. A sense of *finitude,* to use a term from philosopher Telmo Pievani, that can either frighten us or free us.

The king of them all, gravity, is the undisputed master of the Universe, the unrivaled protagonist. It imprisons light, shatters galaxies, transforms into tremors. And its invisible hand is responsible for many fatal attractions.

Except one. As Einstein said, *Gravity cannot be held responsible for people falling in love.*

CHAPTER 4

Bang! And a Story Begins

The Universe Expands

"It is the finest poem written by an American that I have read," wrote Ezra Pound to the editor of the English magazine *Poetry: A Magazine of Verse* in 1915, when recommending the publication of "The Love Song of J. Alfred Prufrock." Pound had met the author, Thomas Stearns Eliot—better known as T. S. Eliot—a year earlier at Oxford, while Eliot was still at the beginning of his career, and became his friend and mentor.

Prufrock is a middle-aged, middle-class man

who is hesitant and anxious. In this imaginary letter to his beloved, the introspective monologue is a disjointed flow of images, questions, and fears in which the protagonist doubts himself and agonizes over whether he will ever find the courage to act, to take a risk, to take the plunge. At least once, at least for love.

> *Do I dare*
> *Disturb the universe?*
> *In a minute there is time*
> *For decisions and revisions which a minute*
> *will reverse.*

Not everyone dares to disturb the Universe. The somewhat unexpected expression seems to allude to a temptation that, whatever the individual nature or pretext, is difficult to resist. And, having been written in 1915, it is almost a premonition. In fact, the final version of Einstein's general theory of relativity, "The Foundation of the General Theory of Relativity," was published a year later, marking a revolution in our understanding of the world and giving rise to modern cosmology.

Prufrock believed it was impossible for his

love to be requited, and this conviction led to his paralysis.

To question the world, for one perspective to expand and encompass others, curiosity and open-mindedness are required. But above all, it is especially how you relate to the perception of the impossible that makes the difference. The courage of an intuition, to put unimaginable questions on the table, and, more than anything else, to believe in your ideas completely. Even at the risk of blanket disapproval.

In hindsight, any revolution in thought seems like an obvious change in direction. Still, it is worth asking ourselves: At the time, would we have enthusiastically embraced or categorically rejected the idea of us being perched on a spherical object where someone, somewhere else, was thus upside down? How would we have reacted to the idea that that sphere rotates on its own axis while simultaneously making large orbits around the Sun? And what about the weird idea that light has to travel to reach our eyes? Even the most enlightened minds, who blaze new trails of knowledge, sometimes cling to the comfort of established beliefs.

Galileo, for example, believed that comets were essentially an atmospheric phenomenon and stubbornly defended the idea that tides were caused by Earth's motion. In his book *Dialogue Concerning the Two Chief World Systems,* he made fun of Kepler, who had "lent his ear and his assent to the Moon's dominion over the waters, to occult properties, and to such puerilities" by accepting the close link between tides and the Moon's movements. And we can only imagine how Lord Kelvin must have felt when the advances of quantum mechanics and the theory of relativity arrived just a few years after his disillusionment with what physics had left to offer.

Einstein was the first to offer a new perspective that led to the rethinking of the Universe. However, it took the intuition and observations of a basketball coach, the work of a priest, a stroke of luck for two telephone engineers, and the nocturnal inspiration of a young man who had just finished his doctorate to *disturb* the Universe more profoundly, to the point of shaking it up, and, through its expansion, giving it a story to tell.

Let us retrace what happened. It all started with a kind of promotion, as we have seen. In physics, speed is described by a simple equation involving space and time. More precisely, velocity correlates the space traveled with the time taken by means of a division: $v = s/t$. When we talk about speed in our everyday language—the car was going *100 miles an hour,* like the title of a famous song—we probably do not realize that we are using an equation. Furthermore, it is a relative velocity; that is, it depends on the observer's reference system. A bag on the seat of that car does not move with respect to its driver. Yet, to a person standing on the pavement watching him speed by, that bag will appear to be moving at 100 mi/h as well.

At the beginning of the 20th century, there was a modification of this perspective. Maxwell's equations, Michelson and Morley's experiment, and Einstein's formulation conclusively established that the speed of light has a unique characteristic. The speed of light in vacuum—186,000 miles per second—is an insurmountable limit and remains unchanged regardless of the observer's reference system.

From being a relative variable, the speed of light was elevated to an absolute quantity, entering the realm of the (few) universal physical constants. This caused an unexpected shift in the rankings. Time and space had now lost that privilege and dropped down the list to become relative.

In other words, once the speed of light has been established as a fixed and insuperable limit, space and time pay the consequences. They must start to adjust to each other in order to keep the speed of light constant in any reference system. They lose their comfortable Newtonian absoluteness. Bound to the speed of light by an arithmetical division, they are forever inextricably linked, and their true inner physical nature emerges.

Einstein's special theory of relativity stemmed from the emergence of this new absolute: the creative power unleashed by constraints.

The next stage of Einstein's reflection was the "generalization" of the theory of relativity to include gravity and accelerated motion. One of his ambitions was to develop a model of a static, spatially curved Universe in which matter was evenly distributed.

But then, a surprise. In a completely unexpected way, and without its author anticipating it, something emerged from those equations that had hitherto been unimaginable, and unimagined. The possibility of a dynamic Universe that, due to the presence of gravity, can expand or contract.

This suggested the Universe had had a starting point, which seemed an extravagant and very far-fetched prospect. Until the publication of the general theory of relativity, the prevailing belief was that the Universe was static, unchanging, and certainly had not suddenly emerged out of nothing. However, if, as Einstein believed, gravity is a force of attraction, then the matter and energy present in the Universe could end up causing it to collapse.

To restore a placidly static Universe and correct this dynamic effect, Einstein resorted to a stratagem, what physicists call a "fudge factor," and arbitrarily introduced the cosmological constant. A kind of vacuum energy that counterbalanced the gravitational effect of energy and matter.

After the publication of "The Foundation of the General Theory of Relativity" in 1916, Einstein's popularity sparked a growing interest in his equations. On the one hand, there was a desire to find new explanations for the structure of the Universe and its contents that were based on experimental data. On the other, there was a fascination with a cosmos that could now be encapsulated in the elegance of mathematical language. A whole world became an equation.

For almost 15 years, Einstein opposed the idea that the Universe was expanding. However, in the interim, new elements, ranging from mathematical proofs to experimental observations, had led to descriptions of a world that was no longer the same.

The first to realize that the solution Einstein had found to render the Universe static was unstable was Alexander Friedmann in 1922. As unstable "as balancing a pencil on its tip," he said.

The Russian physicist developed a new mathematical model consistent with a dynamically evolving Universe. Einstein, however, was not particularly happy about this. His first reaction was to accuse Friedmann of having made a mathematical error. For a long time, he refused to give credit to this solution, until, in 1931, he was forced to recant and admit that Friedmann had been the first to tread the right path.

Meanwhile, the first experimental data had been collected. Despite knowing nothing about Friedmann's calculations, a Belgian priest and astronomer, Georges Lemaître, raised the possibility that the Universe might have had an origin. He did so on the basis not of mathematical hypotheses but of concrete observations, which seemed to confirm its expansion. Lemaître postulated that the Universe originated from a kind of "primeval atom," a very hot, very dense layer of matter.

In 1927, although aware that his ideas might be controversial, he decided to publish his results in a Belgian astronomy journal that had a long tradition but limited circulation, especially internationally. His extraordinary discovery thus went virtually unnoticed at first. But Lemaître persisted. The following year, when he presented his findings at the third general assembly of the International Astronomical Union, the reception was mixed: vivid interest from some and cold skepticism from others. Among the latter was Einstein himself, who, during a private conversation, told Lemaître that while his mathematics was faultless, his interpretation of physics was "abominable." Einstein was wrong.

Lemaître refused to give up. In 1931, he finally wrote up his theory in English in a letter titled "The Beginning of the World from the Point of View of Quantum Theory" for the renowned journal *Nature*. This finally earned him the fame he deserved.

Meanwhile, another astronomer was about

to enter the scene, who would deliver a powerful blow to the prevailing view of the Universe, adding credibility to the brilliant contributions of Friedmann and Lemaître.

Edwin Hubble: a giant in every sense.

Standing 6 feet tall, athletic, he had been a law graduate (but only to please his father), a teacher and basketball coach at a high school in Indiana, and a volunteer soldier during the First World War. Finally, after his father's death, he became an astronomer, pursuing what had always been his greatest passion. In 1919, at the age of 30, he got a position at the Mount Wilson Observatory in California. Armed with the largest, most powerful and innovative telescope of the time, a 2.5-meter (8.2-foot) Hooker telescope, Hubble was to passionately and persistently scan the darkness of Californian nights for the rest of his career, making astonishing discoveries along the way.

One of his first obsessions was Andromeda. A magnificent cloud of light in the sky, thought to be nothing more than a collection of gas and dust in our galaxy. In 1924, Hubble identified some variable stars within the cloud, the Cepheids. He measured their distance from Earth and realized that it was far greater than the known size of our galaxy. These stars were so far away that they had to be beyond its borders. The discovery of Andromeda as a separate galaxy opened up a whole new reality. It was a crucial breakthrough, a real game-changer. Until then, it had been believed that the Milky Way was the whole Universe and now, instead, the Universe had become much more than that.

The following year, Hubble presented his discovery at a conference of the American Astronomical Society and, in 1929, published it in the paper "A Spiral Nebula as a Stellar System, Messier 31," in the *Astrophysical Journal.*

The boundaries of the Universe suddenly widened, and, with this unexpected extension, we once again lost the illusion of being endowed with a privileged position, some sort of uniqueness. Andromeda flanks the Milky Way, and today they are just two bright pinwheels of dust and stars among the approximately 100 billion known galaxies. A hundred billion is the number of galaxies we are able to observe, but their total number is unknown. For all we know, it could even be infinite.

Also in 1929, by comparing observations of the distances of galaxies and data on the light emitted by them obtained by Vesto Slipher and other physicists, Hubble realized another astonishing thing. The farther a galaxy is from us, the faster it appears to be receding. The unmistakable sign of this escape was the redshift in their spectra.

When a light source moves away from an observer, a *redshift* occurs. The wavelengths stretch and the light emitted shifts to the redder part of the electromagnetic wave spectrum. By measuring the dynamics of the galaxies' retreat from each other, Hubble also realized that there is a relationship between the distances of the galaxies and the speed at which they are fleeing: the farther away a galaxy is, the faster it "escapes."

This simple proportional relationship became known as Hubble's law. In mathematical terms, if we call the speed of a galaxy v and its distance from the observer d, we find that all the velocities at which galaxies move away from each other can be calculated by using the law $v = Hd$, where H is a universal constant known as Hubble's constant. If we measure the distance of a galaxy (a complex task), we can calculate the speed of its escape by simply multiplying it by this factor H.

With the now-known value of H, about 70 kilometers per second per megaparsec (or about

43 miles per second per megaparsec), we can establish that galaxies located 100 million light years away are receding at a speed of about 5 million mi/h, while those that are 300 million light years away are fleeing at a speed of about 15 million mi/h. However, this perception of galaxies moving away is not the result of their individual motion. They are not actually fleeing. It is space-time itself that is expanding. It is the very dynamic of the Universe that is pushing them apart.

Space-time expands, and the distance between the galaxies increases. Just as the distance between two dots drawn on the surface of a deflated balloon increases if it is then inflated. No matter where you look from on the balloon, or which galaxy in the Universe you observe the expansion from, you would always get the same results. The galaxies are fleeing. No perspective is privileged. This expanding motion has no true center; each point is as central as any other.

More than three centuries earlier, Giordano Bruno had asserted that "we can affirm that the Universe is all center, or that the center of the Universe is everywhere, and that the circumference is not in any part."

Be careful, though. It is only the distance

between the points that increases. Their dimensions remain unchanged. It would therefore be more accurate to compare the expansion of the Universe to a loaf of olive bread that keeps on rising. The pieces of olives—the galaxies—move away from each other, but their individual sizes do not change. All the objects within a galaxy are held together sufficiently firmly and steadily by other forces, or by more intense gravitational effects, so they do not feel the effect of the expansion. For example, the expansion of the Universe does not result in Earth moving away from the Sun, nor does it elongate the shape of the galaxies.

The Universe is expanding. It is not static and unchanging as was once believed. The energy that fills it causes its expansion and, in that spatially homogeneous motion, it drags the galaxies along with it. And Einstein's general theory of relativity was already out there, ready to provide the mathematical model for the Universe's expansion that could consistently and elegantly explain Hubble's observations.

In 1931, Einstein visited Hubble at Mount Wilson and thanked him for his work. It is said that he described his attempt to defend the idea of a static Universe as "the greatest blunder of my life." In

April of the same year, while giving a lecture at the Prussian Academy of Sciences, Einstein explicitly adopted the model of an expanding Universe and, in 1932, he collaborated with the Dutch theoretical physicist and astronomer Willem de Sitter to propose a cosmological model of a continuously expanding Universe. This model became the generally accepted one until the mid-1990s.

As we have seen, Hubble's law tells us that the expansion of the Universe is governed by the equation $v = Hd$, where H is the so-called Hubble constant. But we also know that speed can also be expressed as distance divided by time, $v = s/t$. Comparing this equation with Hubble's law, we find that H can be represented by the inverse of time, given by the formula $t = 1/H$. By determining the value of H and using its inverse, we obtain the time it has taken the Universe to expand from its beginning to the present day.

There are two complications, however. The first is to find the exact value of H, because it is still extremely difficult to measure the distance to remote galaxies with very high precision.

And this leads us to the second complication. The value of H is not constant but has changed with the evolution of the Universe. Having said that, on the basis of the most precise calculations available we can now say with some confidence that the Universe is approximately 13.8 billion years old.

* * *

Despite the evidence gathered by Hubble, it took a long time before the idea of an expanding Universe (and thus of its dramatic beginning) was accepted in the scientific world. So much so that the first time the expression "Big Bang" was used, it was intended sarcastically. It was the British astronomer Sir Fred Hoyle who came up with it during a radio broadcast on the BBC, while describing this theory on the origin of the Universe that he himself considered dubious:

> These theories were based on the hypothesis that all the matter in the Universe was created in one big bang at a particular time in the remote past. . . . On scientific grounds this big bang hypothesis is much the less palatable of the two. For it is an irrational process that cannot be described in scientific terms. Continuous creation, on the other

> hand, can be represented by precise mathematical equations whose consequences can be compared with observation. On philosophical grounds too I cannot see any good reason for preferring the big bang idea. Indeed it seems to me to be in the philosophical a distinctly unsatisfactory notion, since it puts the basic assumption out of sight where it can never be challenged by a direct appeal to observation.

Hoyle's objection to this model was philosophical as well as scientific. It made no sense, he argued, to discuss the creation of a Universe if there was no preexisting space and time within which the Universe could be formed.

It is not easy to go from the idea of an unchanging and eternal Universe to that of a reality that began to exist "only" 13.8 billion years ago.

The consequences of an expansion are profound. Imagine rewinding a film of the evolution of the Universe as it expands like a balloon. We would observe a gradual contraction, a shrinking volume, becoming smaller and smaller, until it becomes a tiny dot. It would be quite boring to do the same with a static, unchanging Universe, where the "rewind" version would be indistinguishable from the original. Instead, thanks to Hubble's observations, the Universe now has a "once upon a time" from which it can start, a past to tell. In other words, a beginning. Which is what is needed for a story to start. The story of the extraordinary, and still largely mysterious, Universe to which we belong.

* * *

The history of the Universe is also a history of symmetries. It has become gradually apparent, from Galileo's experiments to the formulation of the general theory of relativity, that no place is truly special. No individual observer, regardless of his or her movement, enjoys privileges based on the observer's point of view. Even the theories describing the other three forces—electromagnetic, strong nuclear, and weak nuclear—are based on an abstract but equally compelling set of principles of symmetry. And a theory that is both simple and "neutral" in terms of perspectives of observation is bound to appeal to scientists.

The concept of symmetry is also crucial for

defining time. Its very existence, its flowing forward, seems to be in itself a negation of symmetry.

Yet, the laws of physics that govern the Universe have no preferred direction for the flow of time. They work in the same way whether going forward in time or backward. Like pressing the rewind button when watching a film. If you drop a vase it shatters into a thousand pieces, but if the direction of time were reversed, none of the laws of physics would preclude finding it intact in your hands.

This effect is called T-symmetry, or time-reversal invariance. When scientists sought to look back at the beginning of everything, in order to understand the true nature of space-time, the search for symmetry was an essential perspective that offered various paths through the jungle of possible interpretations.

The first snapshot of the Big Bang theory takes place at approximately 10^{-43} seconds (we are talking about a fraction of a second, a "zero point" with 42 zeros after the decimal point) after the initial instant. In theories of quantum gravity, which envision the unification of the four fundamental forces, this is the moment when gravity is assumed to have separated from the other forces, which instead remained united in a single fusion of electromagnetic, strong nuclear, and weak nuclear interactions. Shortly thereafter, at 10^{-36} seconds after the beginning, strong nuclear force also separates.

It all starts with a primordial soup of extremely high density and temperature. Matter and radiation are one and the same. It is a uniform, isotropic plasma (the same at every point and in every direction) made up of particles interacting with each other at temperatures higher than a trillion trillion degrees. They race in every direction at speeds close to that of light. It is an incredibly hot and energetic Universe, which starts expanding.

Space and time begin with the expansion itself, and the size of the Universe increases exponentially.

At this point, only fundamental particles are present. Atoms, composed of nuclei that have captured electrons and protons, do not yet exist; at these levels of energy and temperature, nothing can stay tethered.

The most dramatic evolution of the Universe occurring during this phase involves both its geometry and temperature. It starts with a steady cooling induced by the extraordinarily rapid expansion of space-time. A cooling that causes the fundamental forces as we know them to gradually emerge. In a tiny fraction of a second, the first to appear is gravity, becoming the predominant force in those very early moments, followed by all the other forces. As the Universe cools down, it no longer has sufficient energy to create new fundamental particles in abundance. As Einstein established in special relativity, for mass to be created, energy must be at least equal to mc^2.

When matter and antimatter collide, they annihilate each other. But fortunately a small amount of matter does somehow survive, which is why, luckily for us, we can be here today to tell this tale.

About 1 second after the Big Bang, the density and temperature of the Universe are such that the ratio of protons to neutrons is "frozen" at 7:1.

In this phase, the first particles to break away from the primordial soup are neutrinos, which emerge from the decay of neutrons. These "elusive" particles are not trapped by anything, as they interact very little with each other or with their surroundings.

Three minutes after the Big Bang, as the expansion continues, the temperature drops from 10^{32} to about 10^9 kelvins. Protons and neutrons bind, forming nuclei that are held together by the strong nuclear force. Neutrons would have been doomed to extinction, as they decay very quickly, but fortunately the density of the Universe at this point is very similar to what we would find inside a star today. Thus, thanks to nuclear fusion, the nuclei of the lighter elements begin to form.

It is amazing to think that when we drink a glass of water—water is made up of two hydrogen atoms and one oxygen atom—the hydrogen we are drinking was almost certainly formed in those first minutes after the Big Bang. A glass of water: a time machine in our hands.

And so, in those early minutes the Universe writes the first letters of the alphabet contained in Dmitri Mendeleev's periodic table.

We are witnessing the process of primordial nucleosynthesis. Helium, deuterium, and lithium nuclei are formed. The cosmic quantities of these elements are set, along with that of hydrogen, whose density has decreased as a result of the formation of helium. If we were to calculate how many of these elements were produced by nucleosynthesis, we would find astonishing results. The mass ratio between hydrogen and helium that we still observe in the Universe today is exactly that established in those first 3 minutes. For every three hydrogen atoms, there is one helium atom.

This is one of the most surprising confirmations of the Big Bang theory. Without primordial nucleosynthesis, the amount of helium we observe in the Universe today could not be explained.

We are at the ABC of the formation of chemical elements. Or rather, the alpha, beta, gamma.

In 1948, one of the pioneers of nucleogenesis, the Ukrainian-born American physicist George Gamow, developed, together with his doctoral student Ralph Alpher, a mathematical model that sought to explain the nuclear processes that would have occurred under the conditions of extreme heat and density after the Big Bang. The result was extraordinary. Using that model, they managed to predict the observed proportions of hydrogen and helium.

In a paper that became a landmark, "The Origin of Chemical Elements," it was explained how, after the Big Bang, subatomic particles fused to form protons and neutrons, and, through subsequent fusions, heavier chemical elements were formed. When submitting the paper for publication in *Physical Review*, Gamow, who had his own particular sense of humor, decided to add a third name, that of his friend Hans Bethe, a renowned physicist who had not contributed to the development of the theory up to that point. Gamow was intrigued by the idea of using the authors' three initials, which were also the first three letters of the Greek alphabet, and so published what became a seminal article, known as the "alpha-beta-gamma paper."

Let us go back to those first few minutes when the Universe is very young. Although it has produced the nuclei of the first light elements, the primordial plasma at this point is still too hot for electrons to be captured to form neutral, stable atoms. Therefore, the matter of the Universe remains an electrically charged fog, which prevents light from passing through it.

It takes 380,000 years for the Universe to cool enough for neutral atoms to be formed. The positively charged nuclei combine with negatively charged electrons in a crucial phase in the history of the Universe, called *recombination*. And it is in this chapter that matter and light begin to exist as separate entities. Each goes its own way, and from this moment on, the Universe becomes observable. No telescope, not even the most powerful one, peering into the depths of the cosmos by means of electromagnetic radiation will ever be able to catch a glimpse of what happened behind that opaque wall, firmly placed 380,000 years after the origin of everything. Although the photons of the cosmic backdrop were formed immediately, about 3 microseconds after the Big Bang, it is only at this point in the evolution of the Universe, and not at its birth, that light can flow freely. *Fiat lux*: let there be light. A new beginning. A new story.

With no more charged matter to scatter them, photons can finally travel freely. The light from the primordial Universe—that is, the radiation released immediately after the beginning of the cosmic expansion—can reach us directly, independently of the other complex developments that await the Universe.

The first light that ever existed still permeates the entire Universe and continues to cool as it expands. It is extraordinary to realize that we can even observe and measure it. These are the same, very first photons that emerged 380,000 years after the Big Bang. Today, the temperature of the cosmic background radiation is just above absolute zero, about 2.73 Kelvin, or about −270 degrees Celsius (roughly −460 degrees Fahrenheit).

After Hubble's observations, detecting this fossil radiation would provide the missing proof, confirming the existence of an expanding Universe.

However, it was not until decades later that the echo of the Big Bang was heard by accident, thanks to an unforeseen circumstance and a bit of luck.

We are in New Jersey, in the first half of the 1960s. Arno Penzias and Robert Wilson, two engineers from Bell Laboratories, were busy investigating the capabilities of radio antennas in space. While taking measurements from a ground-based antenna, they detected a uniform background noise across the microwave range that remained the same, no matter which way the antenna was pointed. A constant noise, present day and night. Only after they had exhausted all reasonable explanations for the origin of that "noise," including the pigeons flying around, did they realize it might be something completely unexpected—a radiation of a different, "extraterrestrial" nature.

Penzias and Wilson had no idea that some 20 years earlier, Gamow himself had been the first to hypothesize the existence of cosmic background radiation. Their discovery reached the ears of Bernard Burke, a radio astronomer at the Carnegie Institution in Washington, D.C., who knew that his colleague Bob Dickie and his team at Princeton were working on cosmic fossil radiation.

As often happens, two distant and completely independent strands of research coincidentally converged, leading to a breakthrough. In 1978, Penzias and Wilson were awarded the Nobel Prize for their discovery: by intercepting that noise, they had provided one of the decisive pieces of evidence in favor of the Big Bang theory.

There remained, however, some loose ends, which were nevertheless important. Even though the Big Bang theory was by then widely accepted, the physicists and astronomers who worked on cosmology in the 1960s and 1970s still had many doubts. These reservations were due in particular to two unresolved issues that the observations had brought to light.

The first, the so-called flatness problem, emerged from the observation that the Universe appears incredibly flat, and the light we see seems to travel along straight lines. This flatness was hard to explain by using the classical version of the Big Bang theory. According to the general theory of relativity, the curvature of the Universe depends on the density of energy, which increases over time. The fact that even now the Universe shows almost no curvature after billions of years of expansion implied that this curvature must have already been small to start with. Very small, in fact, which would have required some fine-tuning, an extremely precise calibration of the initial density of energy. This seemed unlikely.

The second riddle, known as the horizon problem, had to do with observing the Universe on a large scale, as it appears far too homogeneous in terms of physical properties. Regions of the cosmos that are causally disconnected—meaning that they could never have come into contact with each other because the distance separating them is greater than that which light could have traveled in the lifetime of the Universe—have the same average temperature.

Mysteries that remained unresolved. But then along comes the Bang.

* * *

On the night of December 6, 1979, while staying at the Stanford Linear Accelerator Center, the young researcher Alan Guth wrote on the first page of his notebook: "Spectacular realization."

Guth was developing a model of the exponential expansion of the early Universe that could explain the absence of magnetic monopoles—particles with positive or negative magnetic charge—envisioned by the grand unification theory. While reflecting on the then inexplicable flatness of the cosmos observed on scales of hundreds of millions of light years, he realized that his model could offer a physical justification for the extraordinarily homogeneous and isotropic characteristics of the Universe. This "spectacular realization" led to the formulation of a theory that profoundly changed our perspective on the evolution of the Universe and, in particular, on what happened in those first moments after the Big Bang: the theory of cosmic inflation.

According to Guth, shortly after its birth, the Universe experienced an extremely rapid expansion and, in the blink of an eye, its size increased by about a billion trillion times. After 10^{-36} seconds from the beginning of everything, the Universe is thus blasted from microscopic dimensions to cosmic scales. Guth's brilliant insight was to imagine that every small discrepancy in the distribution of matter and radiation in the early Universe would have been completely leveled out by its initial, incredibly rapid expansion.

Guth thus proposed a hypothesis on the first instants after the Big Bang. He described them as a very fast expansion phase in a very short period of time: from 10^{-35} seconds after the birth of the Universe to 10^{-30} seconds. Almost no time at all. According to Guth, in that infinitesimal period of time, the distance separating any two points in the Universe increased by a scale factor of about 10^{26} times. During this inflationary phase, the Universe underwent an accelerated, exponential expansion, multiplying its size in a tiny fraction of a second.

In less time than it takes for a fly to flap its wings, the initial region had expanded from about 10^{-29} meters in diameter to approximately 1 millimeter in diameter by the end of the inflation. All of existence the size of a speck of dust.

Such an extremely rapid expansion would explain the vastness of the Universe and justify dimensions much larger than those consistent with the initial Big Bang model. This was how matter cooled and the curvature of three-dimensional space flattened.

Inflation stretched the fabric of space-time. A single microscopic region, where everything was so close that it communicated through radiation traveling at the speed of light, grew to become the Universe as we know it today, immense and homogeneous.

In the inflationary model, the Universe is vast and flat precisely because it grew so much in its initial phase, just like a specific area on the surface of a balloon appears flatter and flatter as it inflates. This flatness also provides a very precise value for the density of the Universe. A higher density would have resulted in a positive curvature of space, causing it to collapse rapidly, while a lower density would have caused it to expand too rapidly for its structure to form.

Guth's insight offers further, new interpretations. It also explains why localized structures such as stars, galaxies, and galaxy clusters still exist today. According to Guth, matter appears the way it does as a result of tiny irregularities in the primordial Universe, which swelled out of proportion during inflation.

Formulated in 1981, the theory of cosmic inflation was nothing more than conjecture. Guth himself was convinced that it was a very plausible hypothesis but that it would be impossible to provide an empirical demonstration of it. As it turned out, that is not what happened. Instead, today we have many experimental findings supporting this theory.

As Guth said during a 2014 interview with *MIT News*:

> I usually describe inflation as a theory of the "bang" of the Big Bang: It describes the propulsion mechanism that drove the universe into the period of tremendous expansion that we call the Big Bang. In its original form, the Big Bang theory never was a theory of the bang. It said nothing about what banged, why it banged, or what happened before it banged.

In fact the Big Bang did not start with a bang; it did not make any noise. The "bang" was actually an inflation. And this is where Guth's genius comes in: using his own theory, he redefined those first moments that triggered the slow and quiet cosmic evolution that followed. Inflation diluted the initial matter and radiation and caused a sudden drop in temperature. Once inflation ceased, the energy that had driven it transformed into many elementary particles.

The inflation theory was able to answer a number of unresolved questions. Why the Universe contains so much matter, despite having grown from regions of infinitesimally small dimensions. Why it has such a long lifespan. How it managed, despite the enormous amount of matter and energy it contains, to evolve in such a way that it could celebrate 13 billion years and more without collapsing in on itself.

It is amazing how the inflationary paradigm provides an explanation for both the large-scale structure we observe in the Universe—the galaxies, the galaxy clusters—and the small temperature fluctuations in the cosmic background radiation. Because the inflation occurred during the very first moments of the Universe's evolution, when energies were very high, the laws of physics governing its dynamics were those of the infinitely small.

Galaxies originate from quantum fluctuations in the primordial Universe, imprinted as traces in the cosmic background radiation. The tiny ripples of matter, which were stretched to enormous dimensions during the inflation, were the seeds of the subsequent formation of astronomical structures, shaped by gravitational attraction.

In 2013, the Planck satellite detected small fluctuations of cosmic background radiation with incredible precision, charting the entire sky. Their statistical properties perfectly match the assumptions of the inflation theory.

That now yellowed sheet of paper, with the words "Spectacular realization" written on it, is today preserved at the Adler Planetarium in Chicago.

This is the story of the extraordinary evolution of an expanding and changing Universe, of the profound transformation of its geometry and temperature, and, with ever-new chapters, it is also a story of its inherent resilience. As Michel Cassé points out in his article "Rising Stars in the Sky of Knowledge":

> Permanent ebbs and flows between energy and matter, the breaking of symmetries, phase transitions—in short, the evolution of the Universe—cannot, however, conceal the fact that there is some latent eternity: that of calm and solemn laws. The laws of change, at least in a first approximation, do not change.

* * *

Slightly larger than a sheet of printer paper, Adam Elsheimer's *The Flight into Egypt* is a small, intense, and emotional painting. The artist from Frankfurt painted it in Rome in 1609. A few months later, after his premature death, the painting was found hanging on his bedroom wall.

It is nighttime. In that darkness, three sources of light catch your eye. The torch in Joseph's hand as he walks behind Mary and the Child on a donkey. A flickering fire around which shepherds are warming themselves. The reflection of a full Moon on the still lake. Above them, the vastness of a peaceful sky stretches out with its myriad of stars.

However, there is something unusual in that sky. Those bright dots are not just a random distribution of stars in the dark. In that night sky, so extraordinarily poetic and mysterious, something new happens. For the first time in the history of art, the Milky Way appears in a painting, depicted with remarkable accuracy. Elsheimer must have scrutinized the scattered positions of the stars a thousand times, and with patient attention, to achieve such precision. Astonishing accuracy, considering that the publication of Galileo's first astronomical observations came many months later.

Only a hundred years ago, we still believed that the Milky Way *was* the whole Universe. All of existence was thought to be contained within that precise and familiar outline traced by Elsheimer in *The Flight into Egypt*.

After Edwin Hubble observed in 1925 that Andromeda was a galaxy in its own right, the Universe's borders stretched out and began to fill up with more celestial bodies. Today, we know that we and Andromeda are just two of around a hundred billion observable galaxies.

In less than a century, we have understood many things. Our Universe is neither immutable nor eternal: it had a beginning; space-time curves in the presence of energy and matter; it expands.

And, for some years now, we have also known that this expansion is accelerating. The galaxies will gradually become more and more distant from each other. And in billions of years' time, even with the most powerful telescopes, we will not be able to see anything but darkness beyond the Milky Way. The light emitted by distant stars and galaxies will no longer be able to reach us.

The growing distances will create separations that cannot be bridged. This is dictated by the laws of physics, and they cannot be disobeyed.

Looking into deep space, we will see nothing but an endless expanse of silent, dark, seemingly empty stillness. Perhaps we will return once again to imagining the Universe as it was seen before Hubble. Static and immutable. Maybe we will revert to thinking that the Universe is nothing more than our island of light and matter as depicted in that moonlit night in *The Flight into Egypt*.

The unknown is not necessarily a darkness that knowledge gradually illuminates. A mystery can go full circle, taking a turn only to come back around again.

CHAPTER 5

The Reality of Antimatter

The Very First Crime After Creation

THERE IS A STORY ABOUT TWO FRIENDS, who were classmates in high school, talking about their jobs. One of them became a statistician and was working on population trends. He showed a reprint to his former classmate. The reprint started, as usual, with the Gaussian distribution and the statistician explained to his former classmate the meaning of the symbols for the actual population, for the average population, and

> so on. His classmate was a bit incredulous and was not quite sure whether the statistician was pulling his leg. "How can you know that?" was his query. "And what is this symbol here?" "Oh," said the statistician, "this is pi." "What is that?" "The ratio of the circumference of the circle to its diameter." "Well, now you are pushing your joke too far," said the classmate, "surely the population has nothing to do with the circumference of the circle."

This is how a famous 1960 paper titled "The Unreasonable Effectiveness of Mathematics in the Natural Sciences" by Nobel Prize winner Eugene Wigner begins. The reaction to the population-circle association may elicit a smile, but it summarizes a widespread misgiving. What does mathematics have to do with reality? With my world? With me? The instinct to seek out familiarity in language tends to transfer automatically and firmly to mathematics and its symbols. The real or perceived inability to understand something quickly becomes irritation. There is no guide, bridge, or possibility of personal connection. The world that language would disclose remains hidden. It is flat, lacking any depth. Formulas and symbols become a wall behind which you are abandoned.

Certainly, much of the physical world around us can be described. Yet that is only a portion of what mathematics can provide access to. The language of the Universe somehow transcends us, a long thread that reaches far into what is yet to be known.

As a result, unexpected pathways open up through formulas. The giddiness of truths glimpsed, before we are even ready to welcome them. Miracles of connection.

As Wigner wrote, "mathematics plays an unreasonably important role in physics." Mathematical concepts can reveal utterly surprising links. "The miracle of the appropriateness of the language of mathematics for the formulation of the laws of physics.... which we neither understand nor deserve." In his article, Wigner also mentions his genius brother-in-law, Paul Dirac.

Dirac was a brilliant, meticulous, and legendarily solitary scientist. His colleagues and students described him as an eccentric person with a very reserved character, bordering on surliness. He could remain silent for very long periods. At Cambridge, his colleagues coined a specific unit of measurement—the Dirac—to indicate the fewest number of words spoken over the longest period of time. Approximately one word per hour.

As Graham Farmelo reports in his book *The Strangest Man: The Hidden Life of Paul Dirac, Mystic of the Atom,* his closest friends and admirers, including nuclear weapons pioneer J. Robert Oppenheimer, Werner Heisenberg, and Albert Einstein, described him as exceedingly peculiar. As Einstein once said of him, "this balancing on the dizzying path between genius and madness is awful." Einstein himself kept a copy of Dirac's book on quantum mechanics next to his bed and would apparently mutter "Where's my Dirac?" when he came across a particularly thorny problem.

Along with many physicists of his time, Dirac was trying to find a way to bring together quantum mechanics, which applies to matter on microscopic scales, and Einstein's special theory of relativity, which concerns objects moving close to the speed of light.

The matter was of great interest, considering that the smallest particles, such as the electron, travel at such speeds.

Then, in 1928, Dirac succeeded. In fact,

he wrote a single equation—the famous Dirac equation—that made the two theories compatible. It was an immediate success. The equation he developed accurately defined the behavior of the electron and was used to calculate the energy levels that the electron can reach in the hydrogen atom. It was clear from the outset that it was not simply a mathematical formula but a "device" from which a new and exact representation of reality emerged.

Later, in 1933, when Dirac was 31 years old, he was awarded the Nobel Prize, together with Erwin Schrödinger, for his contribution to atomic theory.

The equation did, however, have one oddity. It had two solutions. A bit like the equation $x^2 = 4$, which can be solved by oppositely signed values ($x = 2$ or $x = -2$). One of the solutions to the Dirac equation clearly described the electron, as expected. But there remained the question of what the other could possibly be. Indeed, the equation unequivocally revealed a "twin," identical to the electron but with an improbable negative energy.

Classical physics and common sense could not even contemplate the idea of energies that were not positive. Still less the possibility of a duplication. Mathematics once again caught us by surprise. An electron and a mysterious companion of opposite sign emerged from the equation.

Dirac did not see the new solution as another particle that was independent of the electron. Instead, he interpreted it as another possible state in which the electron itself could be, but instead one defined by a negative energy. However, there was a problem. Physical systems always tend to occupy a state of minimum energy. Just as a ball teetering on the top of a mountain rolls down the slope at the slightest disturbance to come to rest in the valley, so an electron should naturally decay into its negatively charged twin.

In other words, this second solution of the equation even called into question the very existence of the electron itself. It was important to understand what physical interaction was hiding in that result.

Dirac wavered for 3 years before accepting that the negative energy solution could represent something that was actually real. He sensed that this was not just one of those usual mathematical redundancies. His equation had revealed, alongside the known particles, the existence of an opposing reality. The world of antimatter. The equation matched each electron with an *anti-electron,* or *positron,* which was identical in all respects but had an opposite electric charge. This game of mirrors holds true for more than just the electron–it includes everything. Every particle has a corresponding antiparticle.

"God used beautiful mathematics in creating the world," Dirac claimed. Mathematics has an extraordinary inherent power. It can extract threads from the fabric of things we do not yet know about. However, when it comes to understanding the physical world, it is what we observe that counts.

Experimental evidence followed a few years later. Carl D. Anderson was the first to observe a positron, which earned him the Nobel Prize in Physics in 1936.

Anderson had built a cloud chamber that could record the passage of particles flowing through it. The chamber was immersed in a magnetic field that could bend the trajectory of the electrically charged particles traversing it. The curvature of this passage depends on two factors. The first is the charge of the particle: positively charged particles curve to one side, negatively charged ones to the opposite side. The other factor determining the extent of the curvature is the mass of the particle: the lighter it is, the easier it is to curve its trajectory.

In fact, the flow of cosmic rays in Anderson's cloud chamber had left a new imprint. It was similar to the trajectory of an electron but, under the effect of the magnetic field, it seemed to bend following a mirrored curve. This effect could only be explained by considering a particle with a charge that was opposite to that of the electron. There was no doubt. It was a positron: the antiparticle of the electron. Same mass and characteristics as the electron (such as its spin) but with a positive electric charge.

The interesting thing is that Anderson was unaware of Dirac's equation and its two solutions, yet he determined its triumph. "The equation," the British physicist later remarked, "had been smarter than me."

The discovery of the positron was an extraordinary scientific breakthrough, but the question remained as to where it came from. The world is full of electrons, yet finding positrons is by no means simple. Studying the nature of these particles reveals that if an electron and a positron meet, they annihilate each other, turning into very high-energy light radiation in the form of two photons. Positrons are therefore very difficult to find because, whenever they bump into an electron (which are abundant in the world), both vanish in a flash of energy. Indeed, the positron seen by Anderson came from the depths of the Universe. It was the product of a collision between cosmic rays originating from deep space and particles in the atmosphere, generating other particles that are not stably present on our planet.

As encapsulated in Dirac's formula, matter and antimatter always emerge together. They are born in pairs. But if they come into contact with each other, *poof!*, they annihilate each other, leaving behind flashes of pure energy.

In the following decades, it came to be accepted that every particle has an antiparticle, and the fate of annihilation is a now general fact. That said, there are some special particles with no electric charge, such as photons, that have no corresponding antiparticles, because they are their own antiparticles, a bit like $x^2 = 0$, which has only one solution ($x = 0$) and not two.

Matter and antimatter, therefore, cannot coexist. The world we see, capable of containing everything from molecules to complex life-forms, could not endure if it consisted of both matter and antimatter. Our world is made up only of matter. Yet, according to Dirac's theory, matter itself has nothing inherently special about it compared with antimatter. Positrons and antiprotons can create atoms of antimatter just as easily as those of matter.

To test this hypothesis, CERN scientists tried to create antimatter atoms, and they succeeded. In 1996, a handful of antihydrogen atoms were observed for the very first time. Since then, there have been many experiments studying atoms of antimatter. However, it remains very difficult to isolate them because, as mentioned, they vanish upon contact with forms of matter.

The existence and stability of antimatter theoretically open up the possibility of other worlds, other solar systems, or other galaxies consisting solely of antiparticles. After all, antimatter behaves in exactly the same way as matter when it comes to electromagnetic interactions: observing a galaxy through its light radiation enables us to distinguish whether it is made of matter or antimatter.

In reality, physicists are quite convinced of

one thing: the observable Universe is made exclusively of matter. The space between the galaxies of the Universe is not empty, even though it may seem so to the naked eye. It is filled with atoms, such as hydrogen. If there were "antigalaxies," we would see bright flashes in the sky due to the mutual annihilation between matter and antimatter. And so far, this has never been observed.

According to Dirac's equation, matter and antimatter are identical, except for their charge. Although little is yet known about the origin of the Universe, theories concur that in the cosmos that emerged from the Big Bang, there were equal amounts of matter and antimatter.

In a matter of moments, their complete annihilation would have emptied the Universe of all its content.

This cosmic annihilation would still have left a residue, a trail of photons that would have cooled as the expansion continued. But nothing more. No possibility of a world as we know it. How matter managed to prevail over its counterpart remains a mystery. One of the most intriguing in modern physics.

In the beginning, the Universe was very dense and hot, and particles of matter and antimatter interacted, "canceling" each other out all the time. At some point, however, something must have occurred, an imbalance allowing one of the two to survive.

We still do not know the mechanism behind this imbalance.

It would be enough if, for some reason, out of 10 billion pairs of matter and antimatter particles, just a single particle of matter managed to survive the mutual annihilation to explain why the Universe we observe exists.

As Michel Cassé said at a conference: "Au début, il y a la genèse, mais il y a aussi meurtre. La moitié du ciel qui disparaît." (In the beginning, there is genesis, but there is also murder. Half the sky vanishes.)

Where there is matter, there can be no antimatter, beyond an infinitesimal fraction of a second. The hypothesis, therefore, is that a minute portion of matter managed to survive to shape the Universe as we know it.

However, this tiny imbalance is not enough to explain the evolution of matter and the structures of the Universe. In Geneva, the physicists at CERN use the Large Hadron Collider (LHC) to explore the properties of matter and antimatter and search for any subtle differences in their behavior. But, so far, there have been no differences that can shed light on the imbalance that occurred in the first moments of the Universe's life.

On a theoretical level, the Russian nuclear physicist and dissident Andrei Sakharov, who was awarded the Nobel Peace Prize in 1975 for his pro-disarmament and human rights activism, defined the three necessary conditions that must have taken place in the primordial Universe for the imbalance we observe to arise. The first condition is the baryon number violation, meaning that the total number of baryons minus the total number of antibaryons is not conserved. Baryons are particles consisting of quarks, such as the neutron and the proton. There must therefore be some mechanism capable of creating protons without an equal number of antiprotons being formed.

The second condition involves the violation of symmetries: *C symmetry*, which is charge, and *CP symmetry*, which is charge and parity. C symmetry dictates that the laws of physics remain the same even if all the positive charges in the Universe were exchanged for negative charges, and vice versa. CP symmetry is a little more complex and prescribes that the laws of nature must remain the same even if we first apply a change in C symmetry—that is, exchange all the positive charges for negative charges—and then an inversion of spatial coordinates; in other words, swap right for left, bottom for top, and inside for outside. A bit like looking at the Universe through a mirror. Sakharov's second condition for there to be an imbalance between matter and antimatter requires the violation of these symmetries.

In 1964, it was discovered that in the Standard Model of particle physics as we know it, the so-called CP symmetry has in fact already been violated by the decay of a neutral particle called the *K-meson.* This discovery earned James Cronin and Val Fitch the Nobel Prize in 1980. However, although a difference in behavior between matter and antimatter was demonstrated, the violation of CP symmetry in this particular decay was not sufficient to explain the assertion of matter in the primordial Universe.

The final condition is that the Universe must not be in thermal equilibrium.

If Sakharov's three conditions are not met,

the imbalance between matter and antimatter cannot occur.

Sakharov's final condition poses no problem. A Universe like ours that is expanding and cooling is certainly far from thermal equilibrium. However, the other two criteria are more complicated, not least because there is no mechanism that can convert matter into antimatter and vice versa. There is no switching back and forth between these two forms.

There is, however, a different possibility. The invocation of the neutrino. Those elusive particles could be the superheroes that saved the world.

* * *

Antimatter, thanks to its intriguing concept, is quite in vogue. In Dan Brown's sci-fi–religious thriller *Angels and Demons*, an ancient sect called the Illuminati attempts to destroy the Vatican by detonating an antimatter bomb. Perhaps even more fantastical than the bomb itself is the fact that the villains managed to steal a quarter of a gram of antimatter (from CERN), which was needed to trigger the explosion. Tom Hanks probably did not realize that it would take hundreds of millions of years to produce a quarter of a gram of antimatter.

Star Trek's starship *Enterprise* traveled faster than the speed of light by exploiting the mutual annihilation between deuterium and antideuterium atoms. Isaac Asimov's robots could think thanks to a positronic brain, while Marvel's Fantastic Four found the key to the Negative Zone, a parallel world made only of antimatter, by reversing the polarity of all their molecules.

On January 6, 1996, the headline splashed across the front page of the French newspaper *Libération* was FIRST STEPS INTO THE ANTIWORLD, featuring a superhero clinging to a meteorite. The previous September, CERN had indeed created the first antihydrogen atoms, and the article reassured readers that there were no other antimatter particles on Earth.

But it was wrong. Antimatter is something real, even close by. Just take a banana. The potassium atoms in a banana, for example, decay and may sporadically emit positrons.

Antiparticles are consistently observed in cosmic rays or are the result of radioactive decay. Small amounts of antimatter are also produced by lightning during a thunderstorm. However, because they are almost instantly annihilated by matter, natural antiparticles have a very short lifespan.

In positron emission tomography (PET) machines, the presence of tumor cells is assessed by injecting the patient with metabolically active substances labeled with rapidly decaying, positron-emitting radioisotopes. And if one day it were possible to inject a certain number of antiprotons into the tissue where a tumor is located, the annihilations within the tumor cells would release energy and destroy them.

CERN, with its powerful accelerator, is now a veritable antimatter factory. By trapping it, cooling it, and combining it, scientists have managed to produce antihydrogen and study some of its characteristics in detail, such as mass, charge, light spectrum, and behavior in gravity. For now, everything seems to confirm an extreme similarity to the hydrogen we know, with the exception of the charge of its components.

Another instrument "hunting" for antimatter is the Alpha Magnetic Spectrometer, the largest particle detector in space, which has been in orbit on the International Space Station since 2011. One of its scientific objectives is to study the composition of cosmic rays in detail, including detecting the very rare presence of antimatter particles.

There has been a recent discovery that could prove to be very important with respect to the mystery of antimatter. It concerns one of the most elusive particles in the Standard Model: the neutrino, whose existence was hypothesized by Wolfgang Pauli in 1930 to explain beta decay, an important process in nuclear reactions.

It took a couple of decades before it was discovered experimentally in 1956 by physicists Clyde Cowan and Fred Reines. The reason lay in the elusive nature of neutrinos. They are the product of the decay of heavier particles; they have no electric charge, their mass is very small, and they can be measured only indirectly and with poor accuracy. They travel across galaxies undisturbed because they interact only very weakly with matter and so bring with them, depending on their origin, a wealth of unaltered information—information on highly energetic events such as supernova explosions, what happens inside stars (including the Sun) during fusion processes, and the impact of cosmic rays on Earth's atmosphere. Primordial neutrinos, produced in the early stages of the expansion, have since gradually lost energy, becoming practically undetectable.

There is a hypothesis that could make the neutrino the solution to the matter-antimatter problem. Like all known particles, the neutrino has its antiparticle: the antineutrino. According to the Standard Model, the neutrino might be a Dirac particle, that is, the kind of particle described by his famous equation. Theoretically, however, there is an alternative version, the Majorana neutrino. Unlike Dirac's neutrino, the Majorana neutrino is special because it is supposedly the antiparticle of itself. That said, there is currently no evidence supporting the claim that the neutrino behaves as described by Ettore Majorana's rather than Paul Dirac's theory.

One of the experiments chasing this particle is the Cryogenic Underground Observatory for Rare Events (CUORE; *cuore* is also Italian for "heart"). Located in the laboratories of the Italian National Institute for Nuclear Physics under the Gran Sasso mountain in the Apennines, the experiment focuses on observing a particular type of decay that could demonstrate the existence of the Majorana neutrino. This decay is called "neutrino-less double beta decay"—essentially a decay process that occurs without the emission of any type of neutrino. If the scientists working on CUORE were to observe it, it would prove that neutrinos are as described by Majorana, therefore making the matter-antimatter transition possible.

There is a stunning, unique place in the world, hidden in the Kamioka mine in the mountains to the west of Tokyo. Or rather under the mountains, more than half a mile underground. A sort of temple dedicated to antimatter research. A blaze of gold reflected in a brilliant blue.

It is the Super-Kamiokande, or Super-K, as it is often called. It is astonishing to behold. More than 11,000 shimmering gold spheres cover the immense walls, and they are reflected in the stillness of 50,000 tons of pure, transparent water. If you look closely at the photos of the Super-K, you can sometimes even spot someone on a small dinghy, an orange, tiny dot floating in that shining infinity.

The thousands of golden spheres are a multiplicity of electronic eyes (photomultipliers) ready to capture the light produced by the passage of elusive neutrinos as they interact with the pure water. However, this is an extremely rare event, because neutrinos have an extremely low probability of interacting with matter and therefore of being detected. This is why the abundance of water and thus of atoms in the Super-Kamiokande increases the possibility of capturing a neutrino.

The hope is that occasionally a neutrino will collide with one of the many particles in the water tanks and produce light radiation, which will then be detected by the photomultiplier tubes and measured.

This experiment studies an attribute of neutrinos called "flavor." Flavor is a property of particles that, just like charge and mass, defines their characteristics and how they interact with other fundamental particles. Usually, fundamental particles do not alter their properties very easily. An electron with a positive charge, for example, will never change its charge. However, neutrinos have a special characteristic. They can come in three "flavors," known as the muon, electron, and tau. As they travel, they can "oscillate," transforming into different flavors. Like the flavor of an ice cream that can change, going from chocolate to a bit of strawberry or becoming vanilla.

Thanks to Super-Kamiokande, the probability of a neutrino switching from one flavor to another has been studied more thoroughly. And therein lies the surprise. While looking for differences in the way neutrinos and antineutrinos change flavor, it has been observed that neutrinos seem to be much more likely to do so than antineutrinos. This discovery is huge. Neutrinos and antineutrinos are not necessarily a reflection of each other but instead behave differently. In recognition of this discovery, on April 15, 2020, the journal *Nature* featured Super-K on its cover with the jubilant headline: THE MIRROR CRACK'D.

Waiting for the Next Revolution

The narrating voices that, entwined or in juxtaposition, tell us about the Universe are only two. Light and gravity.

In the story that emerges, the invisible holds sway. What we can observe is simply an exception.

Let us begin with that small portion of the world that we can see with our own eyes. The light we are able to perceive, the visible radiation, comprises only an infinitesimal fraction of the immense range of wavelengths in the electromagnetic spectrum.

A tiny sliver of radiation through which we observe the world and try to peek further afield. The portion of the spectrum we have access to is so limited that we are immersed, more or less consciously, in a cosmic reality that does not manifest itself, that we cannot see. The darkness that frightens is not intrinsic to night. It is the consequence of our blindness, our inability to perceive more.

Yet it is within the visible radiation, this slice of electromagnetic waves, that we find the colors, or rather the perception of colors, that the eye captures and the brain processes. Violet and red are the two extremes, the boundaries of visible radiation, edges within which white light breaks down into the shades of the entire rainbow. Beyond that spectrum bubble granted to us, everything is gray.

The invisible world disclosed by light is, therefore, of a different nature from that concealed or revealed by gravity. Its origin is subjective. It is invisible because it is inaccessible to our eyes. In order to detect and observe it, we have to rely on increasingly sophisticated technologies. Telescopes, detectors, and instruments, which reveal realities hidden in remote wavelengths.

Some examples of what we can observe only with their help: Gamma-ray bursts—the most energetic waves in existence—which are triggered by extreme events, such as matter-antimatter annihilation or supernova explosions. The "microwave sky" that conceals the first light that ever existed, the first photons that emerged after the Big Bang, a cosmic background of primordial light that permeates the Universe and accompanies it in its expansion. Ancient galaxies, forming planets, and a thousand other wonders, which are all revealed by peering with the help of technology into that ice-cold infrared radiation.

Thus, as we continue to scrutinize, map, probe, and measure, the Universe gradually reveals its wonders. A Universe that is rich, immensely rich, sometimes violent, astonishing, and constantly transforming. About a hundred billion observable galaxies, each with hundreds of billions of stars, an abundance of planets, comets, meteorites, and asteroids streaking through the darkness, and then interstellar gas and dust filling infinite distances. Yet all these marvels, all these cosmic objects made of atoms and emitting light through the whole spectrum of radiation, represent only 5 percent of what exists in the cosmos. "The siege of the visible—the power of the invisible," as in the title of a poem by Valerio Magrelli. The missing portion is nearly the entirety, an immense 95 percent, that does not emit any light and is made of mysterious "stuff" that does not interact with photons.

* * *

What is missing and hidden, that overwhelming 95 percent, only gravity can tell. It is the "dark" part of the story. That of a mysterious cosmos, permeated by an obscure form of *matter* that is not made of atoms, and which expands and accelerates under the influence of an equally obscure energy. And that is not all. It may even be part of a multiverse that could be hiding additional dimensions within the folds of space-time.

If we look solely at the matter that fills the Universe, there is a portion that is an absolute mystery because it does not emit any form of light. The first indications of this invisible presence date back to the early 1930s. Swiss astronomer Fritz Zwicky had recently started working with Edwin Hubble, having moved to the United States for a research assignment at Caltech in 1925. It did not take him long to discover an inexplicable contradiction in the observed mass of the Coma Supercluster some 320 million light years from Earth. Focusing on this bright and rich cluster of thousands of mostly elliptical, old, and red galaxies, he measured their mass using two different methods.

One method took into account their luminosity while the other used, for the first time, a theorem borrowed from statistical mechanics: the virial theorem. This theorem enabled him to deduce the mass of any galaxy on the basis of the "peculiar" velocity of the stars therein, or in other words, their velocity with respect to the reference system of the galaxy itself.

Comparing the results, he found that something was clearly wrong.

The speed at which the stars were moving was much greater than that consistent with the amount of visible matter observed. At those speeds, it should not have been possible to keep the cluster as compact and stable. And so the first doubts arose: the possibility of a presence that light does not reveal, but which nevertheless "counts" by intervening gravitationally in the delicate equilibrium of the Universe.

The hypothesis put forward was quite revolutionary. The galaxies, as well as the space between them, seem to be permeated by a new and incredibly peculiar type of matter, dubbed *dunkle Mate-*

rie, or dark matter. "Dark" because it neither emits nor absorbs any form of electromagnetic radiation, making it invisible, and because it does not interact in any way with ordinary matter, except through the gravitational attraction it exerts or through other very weak interactions.

The hypothesis was published in 1933 in the journal of the Swiss Physical Society and was met with skepticism, while the discrepancy in the results from Hubble was put down to technicalities. Today we know that *dunkle Materie* is not only real but also predominates. After Zwicky, the issue was largely ignored for a few decades. Then, 40 years later, a new measurement emerged, leading to the second anomaly.

The astronomer Vera Rubin—the first woman permitted, in the mid-1960s, to make observations using the powerful telescopes of the Palomar Observatory in California—and her colleague Kent Ford studied the rotational velocity of stars in spiral galaxies, starting from the center of the galaxies and gradually moving outward. Kepler's third law, Newton's theory of gravity, as well as general relativity and the existing evidence of what happens in the Solar System, suggested that their speed of rotation should decrease as they moved from the center of the galaxy to the periphery. Yet this did not seem to happen. What Rubin and Ford observed was surprising. As the stars moved away from the center, their speed remained constant from a certain point onward. They showed no sign of slowing down as expected. In a paper published in 1980, the two scientists illustrated the velocity curve of the stars of many galaxies, including our neighbor Andromeda, revealing that their rotational velocity did not depend on their distance from the center. As with Zwicky's observations in the 1930s, the only way to explain this "oddity" seemed to be to accept that every galaxy was immersed in a dark but gravitationally active element.

The Swiss physicist's ingenious insight thus made a comeback and was finally fully accepted in the 1980s, after his death. Since then, the most diverse hypotheses on the origin of dark matter have continued to emerge. But the answer, the perfect model, is yet to come. And this is thrilling, because it is in such quests that science flourishes—the irresistible seduction of what eludes us, the lure of a glimpsed answer.

What we do know about dark matter, however, is not insignificant. Today's findings suggest that, like an ethereal tapestry, dark matter envelops galaxies, giving them the shapes we see, extends into intergalactic space, and accounts for more than five-sixths of the matter in the entire Universe, thus playing a key role in its formation.

Unlike ordinary matter—which can dissipate its energy by emitting electromagnetic waves, like a star radiating light—dark matter does not radiate. Therefore, by its very nature it retains more of its energy and is more stable. That is why it tends to take longer to create more durable, compact structures. Its gravitational influence enabled ordinary matter to densify quickly enough to survive the expansion of the Universe during the formation of primordial structures, resulting in huge aggregates such as superclusters of galaxies.

Today, dark matter is organized into clusters and filaments that envelop and embrace that ordinary, luminous matter that makes up everything we have direct experience of in the Universe. The Milky Way, other galaxies, and clusters of galaxies have taken shape within concentrations of dark matter, cosmic cradles that locally counteract the expansion of the Universe, preventing galaxies from dispersing.

The scientific debate about the nature of this new type of matter is still ongoing, and the exploration of possible models is in the hands of particle physics. The Standard Model of elementary particle physics contains, as of today, 37 fundamental particles: a photon (mediating the electromagnetic force), eight gluons (mediating the strong force), two W bosons and a Z boson (mediating the weak force), six leptons (the electron, the muon, and the tau, plus their "neutrino partners"), 18 quarks, and a Higgs particle; to this should be added the graviton, the as-yet undetected mediator of the force of gravity. These particles are bound together by a complex network of finely tuned interactions.

Introducing new particles into the system is by no means straightforward, but some speculative theoretical models inherently contain certain particles that might have the right characteristics to describe dark matter.

The Standard Cosmological Model, which today contains all the ingredients to explain the formation and evolution of cosmic structures, requires that dark matter be composed of "cold," low-energy particles that interact very weakly with each other and with ordinary matter. It is a compelling model that clearly describes the structure of the Universe from the cosmological to the galactic scale. However, there are astronomical phenomena that cannot be explained if we assume that dark matter is composed solely of particles in the classical sense of the term, and which instead seem to require a different formulation of the law of gravity. The resulting set of theories is referred to by the acronym MOND (modified Newtonian dynamics) because they are based on the introduction of modifications to the second fundamental principle of dynamics, which says that the force a body is subjected to is equal to its mass times its acceleration. In MOND theories, the relationship between force, mass, and acceleration is more complex and would resolve, for example, the observed discrepancy in the rotational speed of stars in galaxies without the need for dark matter. But no convincing formulation has yet been found, leaving these theories incomplete, and still highly debatable.

* * *

Several studies have attempted to formulate a speculative hypothesis that reconciles the two views: dark matter in the form of particles and modified theories of gravity. One of them, proposed by the two physicists Lasha Berezhiani and Justin Khoury in 2015, is drawn from one of the most fascinating phenomena in physics: phase transition. Water turning into ice, for example.

These phase transitions can be described as drastic mutations in the behavior of a physical system, requiring a completely different mathematical description. When water transforms from a liquid to a gaseous state, for instance, its behavior changes radically, as do the equations used to describe it. In the case of dark matter, the two physicists hypothesized that within the galaxies, it might behave like a so-called superfluid, a system with no internal friction characterized by the collective behavior of its constituent particles. Thanks to quantum effects, these particles can interact even when separated by great distances. Superfluid dark matter may have formed as a result of a phase transition, a shift from a description in terms of simple particles to one in terms of a field interacting with gravity.

Although still speculative, the superfluid theory is based on a number of observations and allows us to think of dark matter as a gas that is made up of particles and which permeates the Universe and condenses into droplets around galaxies, much like water vapor condenses in the air because of the presence of atmospheric dust.

Regardless of the validity of this theory, the possibility of dark matter turning into invisible rain is a beautiful, highly poetic image.

* * *

In 1917, Einstein made a "mistake" that he recognized only some time later. The equations describing the Universe based on the general theory of relativity suggested a dynamic Universe. The presence of matter and radiation in the equations introduced an element of instability and the prospect of expansion or contraction, which Einstein had not anticipated. To restore a static and immutable Universe, he decided to add a mathematical construct to his equations, denoted by the symbol Λ and known as the cosmological constant. This constant modified the geometric formulation of the equations and allowed for a static solution. The Big Bang theory of an expanding Universe was established over time on the basis of Lemaître and Edwin Hubble's observations of receding galaxies in the late 1920s, the discovery of the cosmic background radiation in the 1960s, and Alan Guth's "spectacular realization" leading to the inflation theory in the late 1970s.

However, it was still believed that the expansion was slowing down due to the gravitational attraction of the galaxies. The cosmological constant had therefore become redundant.

In 1998, two different teams of astronomers, through independent observations but in perfect synchronicity, arrived at a startling conclusion. The expansion of the Universe is accelerating. Physicists Saul Perlmutter of the Supernova Cosmology Project and Adam Riess and Brian Schmidt of the High-Z Supernova Search Team realized this when they noticed the distance to some supernovas was greater than expected. This was only possible if the Universe was expanding *faster* than expected—an incredible discovery, worthy of a Nobel Prize, which the three scientists received in 2011. A reason for this could be the presence of some sort of dark energy, an expression first used by cosmologist Michael Turner in the late 1990s. And with this new mystery, our understanding of the Universe is once again irreversibly transformed.

The only forward thrust known to scientists was that which the Universe had received at the moment of its birth. However, the impetus of the Big Bang alone could not account for the observation of a perpetual acceleration. The measurements could be explained only through a different mechanism, imagining a sort of mysterious "cosmic engine" (dark energy) that, working against gravity, expands space-time at an increasingly rapid rate. An entity, invisible and alien, acting as an antagonist to the attracting effect of matter. Something that pushes forward, while gravity pulls back.

Through the observations of Perlmutter, Riess, and Schmidt, the evidence of an accelerated expansion of the Universe brought back into play that Λ term that Einstein had "forced" into his equations more than 80 years earlier in order to restore a static Universe. This cosmological constant, initially a mathematical convenience, now acquired an active physical role, alongside matter and energy, in shaping the geometry of space-time. But with a difference. In Einstein's equations, matter and energy always have positive values, whereas the cosmological constant can take on either a positive or a negative sign, with opposite dynamic effects. A negative cosmological constant exerts an attracting function, which, when added to the equation of general relativity, acts to slow down or balance the expanding Universe. With a positive sign, its effect becomes repulsive, a sort of antigravity that manifests itself in an accelerated expansion. A negative or positive Λ, and the dynamic of the Universe becomes a whole different story.

The challenge was now to understand the true nature of the cosmological constant: a parameter that has today become essential in cosmological models, and which acts as a constant repulsive force whose value changes depending on the theoretical model considered. And it also seems to depend, at least in part, on a counterintuitive, but fascinating, concept: the energy of a vacuum. It might be surprising, but in space, as with darkness, emptiness does not mean absence. For very brief intervals, particles and antiparticles are created and destroyed because of quantum field fluctuations. Usually, these fluctuations cancel each other out, and in fact appear not to even exist. On a cosmological scale, however, all this commotion leaves a trace: it produces a mean energy density, uniformly permeating the fabric of "empty" space-time. The cosmological constant can be regarded as a *zero-point energy*—that is, the resulting energy of the fundamental state of all possible energy states of all particles and fields in the Universe.

Physicists calculated its theoretical value on the basis of the laws of quantum physics only to discover a new surprise.

The value of the cosmological constant that emerges from mathematical calculations is infinitely greater than that measured experimentally by observing the acceleration of the galaxies. A whopping 120 orders of magnitude difference—a 10 followed by 119 zeros. A discrepancy that cosmologists have called the "vacuum catastrophe," or the "cosmological constant problem." It remains an unresolved issue that has led to the formulation of theories that introduce new quantum fields. These fields, with their fluctuations, could offset the zero-point energy of the vacuum, resulting in the tiny, yet positive, value of the cosmological constant observed experimentally.

These are all just speculations for now.

Yet, regardless of its origin, whether it is the result of vacuum fluctuations or some unknown new quantum field, dark energy seems to be a kind of negative pressure, one that drives expansion. A form of energy that permeates the Universe, that does not clump, that maintains the same density everywhere in space and time, and that does not dilute as the Universe expands. It is this invariability over time that justifies the designation of the cosmological constant as the simplest form of dark energy.

Dark energy is not transported by particles or matter, and it has been present in the Universe since the beginning of time. But it had to wait a long time to assert itself. At the dawn of the Universe, much of the energy was transported by radiation.

However, as the Universe expanded, radiation diluted more quickly than matter, gradually taking a back seat. The "baton" was then passed to matter, which began to be the predominant component of energy. Much later in the evolutionary process, dark energy, which does not dilute, took over. Today, dark energy is estimated to account for about 69 percent of the total density of energy of the Universe.

The accelerator that has fueled the expansion of space-time since the very first moments of the cosmos is engaged in an incessant arm wrestle with the aggregating action of ordinary matter and energy. A tug-of-war on which the fate of the Universe depends.

Today, however, it is not the existence of dark energy that is the true mystery: quantum mechanics and the general theory of relativity not only substantiate it but even predict its existence and physical consequences. The question that still remains open is why its measured density is so low. And what this might conceal.

* * *

In a letter to his young son, the nearly 80-year-old Reverend John Ames, protagonist of Marilynne Robinson's *Gilead*, asks:

> When did you figure out about the pantry? That's where we always put anything we don't want you getting into. Now that I think about it, half the things in that pantry were always there so one or another of us wouldn't get into them.

These are the elements of the Universe. Ordinary matter and, "in the pantry," dark matter and dark energy. According to the latest estimates, 5 percent of the Universe is composed of ordinary matter, atoms that emit light. Beyond that, there is only darkness. About 26 percent of this imaginary cosmic pie consists of dark matter, while dark energy is the dominant portion, an impressive 69 percent of this vastness. There is another small piece that contributes to the composition of the Universe: neutrinos. An elusive little voice, which stands alongside those of gravity and light, adding to the grand narrative.

The dark components of the cosmos reveal themselves through the effects they produce, leaving hints here and there, but there is no definitive proof of their nature as yet. That is why it is important to be able to observe a vast portion of the Universe in order to understand their large-scale impact.

By assigning precise space-time coordinates to all the observable galaxies created in the past 10 billion years, the ESA's Euclid space telescope will attempt to reconstruct a three-dimensional map of the Universe in order to understand how the dark components influence its shape and structure. Other missions peering into the mysteries of the dark Universe will add to Euclid's readings, among them NASA's WFIRST telescope and the Dark Energy Survey (DES) in the United States, and the Vera Rubin Telescope (or LSST) in Chile. The quantity and quality of data from the Euclid space telescope will make it possible to test eventual modifications to the general theory of relativity—and to see whether, by adapting the law of gravity to the vast cosmological scales, there really is a need to find an alternative explanation for what we interpret as dark energy.

Understanding the nature of both dark energy and dark matter is an exciting and complex challenge, involving indirect findings and hidden evidence. The most powerful image of the Universe is, of course, the first rays of light to emerge from the darkness some 380,000 years after the Big Bang: the cosmic microwave background. Observing primordial gravitational waves may one day tell us more about what happened before that time, in the early chapters of the history of a hot, dense, and still opaque cosmos.

Alan Guth's inflationary models neatly reconcile some inconsistencies between what has been observed and the Big Bang theory. Yet there is a piece missing. They do not include a mechanism that could put an end to the very rapid inflationary expansion. In the late 1980s, some scientists began to wonder what would happen if this expansion never ended.

The calculations pointed to an intriguing hypothesis, the idea that the Universe is actually undergoing an "eternal" inflation, which may have occurred multiple times, leading to the formation of countless "pockets" of space-time filled with matter and radiation. Real, independent universes, bubbles of stillness, sheltered from the overwhelming inflationary expansion, suspended in an ever-expanding multiverse. Separate worlds, therefore inaccessible, governed by their own physical laws, with specific fundamental constants, and filled with particles that may have been formed by mechanisms different from those we know. In this new perspective of eternal inflation, the Universe itself is just one of these bubbles, one version of the possible and random configurations of matter and energy generated by inflation.

The question of why the fundamental constants of the Universe have such extremely precise values—the only ones compatible with our existence—acquires a different meaning. The seduction of an anthropocentric model has been overcome. And the question of why dark energy has such a peculiar value has been resolved: each pocket/bubble could, in principle, possess fundamental constants and a density of dark energy that differ from those of the others. Our bubble is simply the only one we inhabit, the one that has the right parameters and amount of energy to ensure the emergence of a Universe with the characteristics we observe.

A new paradigm that shatters the last illusion of some "centrality" about the world we inhabit. Our Universe is neither special nor unique.

As Michel Cassé wrote:

> Science is a long battle against geocentrism and anthropocentrism, a progressive decentralization that produces narcissistic pain in some, ecstasy and liberation in others.

The idea of a multiverse populated by many "bubble universes" governed by different physical laws, although bizarre in principle, could, according to some, be substantiated by a rather important element in the puzzle of contemporary theoretical physics: string theory.

Let us begin with matter. The Standard Model

theory provides a description of all the constituents of matter in the form of elementary particles and their fundamental interactions. If we examine matter as if we were unpacking a Russian doll, we would find the atom, which contains a nucleus, that in turn contains neutrons and protons. Delving deeper, we would find the last, the smallest, building blocks, the quarks. These particles have no intrinsic mass; instead, they acquire it through interaction with a fleeting particle, the Higgs boson, which was observed in 2012 thanks to CERN's LHC particle accelerator.

Today, string theory is one of the most widely accepted theories (albeit with reservations and limitations) for explaining the existence of all the particles included in the Standard Model. But it is more than that. It also provides an answer to one of the central questions of modern physics: how to integrate the principles of general relativity with those of quantum mechanics and trace the microscopic origin of all fundamental forces, including gravity.

According to string theory, the ultimate constituents of matter are not "point-like," as we imagine quarks and leptons to be. There is yet another layer. Penetrating deeper, we find something else: the fundamental elements of matter, filaments of energy that, by vibrating at specific frequencies, generate the particles we observe. Like the strings of a violin that, depending on the vibration they are subjected to, produce a wide variety of notes.

Strings can come in two forms: closed or open. Closed strings are free to propagate in every direction in space and describe the behavior of gravitons (the particles mediating gravity, and which have not yet been directly observed). Open strings, on the other hand, exist only in the four dimensions of space-time and instead describe the behavior of the mediators of the fundamental forces, such as photons and gluons. These open strings are connected at their ends to multidimensional membranes called D-branes (Dirichlet branes). The tethering points of the strings attached to the D-brane can be thought of as point particles. We could also think of our Universe as a three-dimensional D-brane stack with many strings attached to it that appear to us as particles.

Unlike general relativity, string theory requires more than three spatial dimensions to be mathematically consistent. Some formulations envision

25 spatial dimensions, others 9 or 10. The extra dimensions that were present at the beginning of the Universe may have compacted or curled up over time to sizes that are imperceptible to the human senses. Their existence could be proved only by experiments involving extremely high energies.

Extra dimensions. Rolled-up dimensions. Incredible, yet possible. We "Newtonian earthlings" perceive space only in a three-dimensional way, assigning height, length, and depth to any object. In order to accept how it is possible to live in a world that we perceive as three-dimensional, but which in reality consists of additional and hidden spatial dimensions, we must accept the limitation that our senses impose on our ability to perceive realities beyond us.

Imagine being able to see only what is happening on a tabletop, a purely two-dimensional view of the world, which excludes the possibility of perceiving the vertical third dimension. Two chips are thrown hard at each other on the table. If they collide with sufficient energy, one of the chips will bounce out of our observation plane. In our view, which is limited to the table, we will see it disappear, as if it no longer exists, when in fact it has only ended up in a spatial dimension (the vertical one) that we do not have a direct view of, but which does nevertheless exist. A similar effect could be occurring with the extra spatial dimensions predicted by string theory.

The existence of additional dimensions also provides an interesting solution to the seemingly inexplicable "hierarchy problem." As we have seen, of the four fundamental forces, that of gravity is by far the weakest—as much as 40 orders of magnitude weaker than strong nuclear force. Because closed strings, which mediate the force of gravity, can move in any direction, part of the intensity of the gravitational field would be dissipated within the compacted dimensions. Gravity becomes like a spring of water dispersed into a thousand rivulets. Other forces, such as electromagnetic force, are mediated by open strings and are thus restricted to propagating only in the space-time dimensions we observe, without dispersing into the extra dimensions.

It is not easy to envision large, easy-to-see dimensions alongside tiny, crumpled dimensions, so small that, even if we were surrounded by them, we would not perceive them.

In his TED Talk "Making Sense of String Theory," Brian Green gives this example. Imagine a cable in Manhattan supporting a suspended traffic light. The cable looks one-dimensional from a distance, yet it has a certain thickness, which is made up of many interwoven fibers that are difficult to see from afar. But if we zoom in and adopt the perspective of a tiny ant walking along that cable, we will be able to see all the dimensions—not only the length, but also the clockwise and counterclockwise directions one can travel along the cable fiber.

We can therefore imagine that we experience the large dimensions, but that there could be additional ones that are rolled up, like the circular fibers of that cable, whose twists and turns remain invisible to us.

Dimensions that are rolled up or "compactified" in different ways produce the various interactions between fundamental particles. In our Universe, as in others, the laws and constants of physics emerge from the way these strings are rolled up.

String theories envision several additional spatial dimensions, which can be folded back on themselves in a multitude of ways. The number of possible "compactifications" capable of producing plausible physical laws is estimated to be equal to or greater than 10 raised to the power of 500—a 10 followed by 499 zeros of possible universes. This is a scenario known as string landscape. In this scenario, we can imagine that each different type of string theory describes one of the universes created during eternal inflation: the extra dimensions can be rolled up differently in different bubbles, resulting in universes with different physical constants.

Cosmic inflation, therefore, may be occurring continuously in distant corners of the cosmos, creating worlds that are part of a vast "multiverse" in which they are all embedded. Our Universe would be just one of those bubbles floating in a cosmic foam.

* * *

What the fate of our Universe will be is difficult to say.

The most widely accepted theory at the moment sees us heading toward a "Big Freeze." The estimated amount of dark energy seems to be high enough to guarantee an indefinite expansion of the Universe, but not enough to overcome the gravitational attraction that holds galaxies together. Inevitably, when every star has completed its life cycle, the Universe will be dominated by black holes that, as Stephen Hawking hypothesized, will slowly evaporate by emitting radiation due to quantum effects.

Eventually there will be a heat death, a state in which energy will be evenly distributed across a Universe in perfect equilibrium. As Bob Dylan sings in "Not Dark Yet," we are not there yet, but some day we will be.

Among other possible scenarios is the "Big Bounce." In this model, the Universe seems to breathe in and out: each phase of expansion would be followed by a phase of contraction, a "Big Crunch," where matter condenses, temperatures rise, and fundamental particles move independently of each other, freeing themselves from nuclei, until the level of concentration is high enough to trigger a new expansion and a new creation.

If the cosmological constant were not truly constant, a violent rending, or "Big Rip," could also occur, which would tear apart all matter and even space-time itself.

In one of his most famous quotes, Blaise Pascal sums up a powerful feeling triggered by looking up at the skies: “The eternal silence of these infinite spaces terrifies me.” Indeed, it is difficult to look up and remain indifferent.

Dark matter. Dark energy. Antimatter. Multiverses. Quantum gravity. Mysteries. Destiny.

We are navigating through shifting reference points, leaving behind old convictions, encountering surprising observations, and witnessing evolving technologies. Impatiently awaiting new measurements, unprecedented questions, and fresh insights. It is a yearning that never ends. It is science. The greatest of all adventures.

Acknowledgments

There are many people and friends to whom I am grateful for their generosity and patience in accompanying me on this adventure. I would like to mention some of them. Jeff Israely, for being the first to encourage me to put the stories I love to tell into a book. Michel Cassé, astrophysicist and poet, for our precious conversations, which have been an extraordinary source of inspiration. My lifelong friend, Massimo Bianchi, a theoretical physicist who never shies away from a question and from whom I continue to learn. Paolo de Bernardis, an experimental astrophysicist, for his thoughtful advice, and to whom I owe my fascination with laboratories, helium-3, and fossil radiation. Professor Robert Jantzen for his precious advice and friendship. Corrado di Giulio, the curious and demanding first reader of this book. Bianca Cardi, for her philosophical clarifications. The young particle physicist Nicolò Foppiani, for his insights. My colleagues Paolo Ferri and Fabio Favata, and Nicola Curzio, who quizzed me on the resilience of the Universe and knew how to listen.

And my thanks, with love, to Stefano.

Bibliography

Alpher, Ralph, Hans Bethe, and George Gamow. "The Origin of Chemical Elements." *Physical Review* 73, no. 7 (April 1948): 803–4.

Barnes, Julian. *Levels of Life*. New York: Vintage, 2013.

Bradt, Steve. "3 Questions: Alan Guth on New Insights into the 'Big Bang.'" *MIT News*, March 20, 2014. https://news.mit.edu/3-q-alan-guth-on-new-insights-into-the-big-bang.

Buonomano, Dean. *Your Brain Is a Time Machine*. New York: Norton, 2017.

Cassé, Michel. *Énergie noire, matière noire*. Paris: Odile Jacob, 2004.

Cassé, Michel. "Preface." In *Stellar Alchemy: The Celestial Origin of Atoms*, translated by Stephen Lyle, xi. Cambridge: Cambridge University Press, 2003.

Cassé, Michel, and Edgar Morin. *Enfants du ciel: Entre vide, lumière, matière*. Paris: Odile Jacob, 2003.

Cooperrider, Kensy, and Rafael Núñez. "How We Make Sense of Time." *Scientific American* 27, no. 6 (November 2016): 38–43.

Einstein, Albert. *Relativity: The Special and the General Theory.* London: Penguin, 2006.

Eliot, T. S. "The Hollow Men." In *The Waste Land and Other Poems,* 81–96. New York: Vintage, 2021.

Eliot, T. S. "The Love Song of J. Alfred Prufrock." In *The Waste Land and Other Poems,* 5–10. New York: Vintage, 2021.

Farmelo, Graham. *The Strangest Man: The Hidden Life of Paul Dirac, Mystic of the Atom.* New York: Basic, 2009.

Galilei, Galileo. *Dialogue Concerning the Two Chief World Systems.* Edited by Stephen Jay Gould. Translated by Drake Stillman. New York: Modern Library, 2001.

Greene, Brian. *The Fabric of the Cosmos: Space, Time, and the Texture of Reality.* New York: Vintage, 2004.

Greene, Brian. *Until the End of Time: Mind, Matter, and Our Search for Meaning in an Evolving Universe.* New York: Vintage, 2020.

Hubble, Edwin. "A Spiral Nebula as a Stellar System, Messier 31." *Astrophysical Journal* 69, no. 6 (March 1929): 103–58.

Hugo, Victor. *Promontorium Somnii.* Edited by René Journet and Guy Robert. Paris: Les Belles Lettres, 1961.

Impey, Chris. *Einstein's Monsters: The Life and Times of Black Holes.* New York: Norton, 2018.

Kepler, Johannes. *Astronomia Nova.* Heidelberg: Vögelin, 1609.

La Capria, Raffaele. *Ferito a morte.* Milan: Mondadori, 2021.

Leibniz, Gottfried Wilhelm. *Discourse on Metaphysics and Other Writings.* Edited by Peter Loptson. Ontario: Broadview Press, 2012.

Lemaître, Georges. "The Beginning of the World from the Point of View of Quantum Theory." *Nature* 127, no. 3210 (May 1931): 706.

Leopardi, Giacomo. *Storia dell'Astronomia.* Roma: Edizione dell'Altana, 2011.

McKeon, Richard, ed. *The Basic Works of Aristotle.* New York: Modern Library, 2001.

Montale, Eugenio. "The Balcony." In *The Collected Poems of Eugenio Montale.* Edited by Rosanna Warren. Translated by William Arrowsmith. New York: Norton, 2012.

Newton, Isaac. *Opticks.* London: Sam. Smith and Benj. Walford, 1704.

Newton, Isaac. *Principî matematici della filosofia naturale.* Torino: Einaudi, 2018.

Pascal, Blaise. *Pensées.* Translated by A. J. Krailsheimer. London: Penguin, 1995.

Popova, Maria. *Figuring.* New York: Vintage, 2020.

Randall, Lisa. *Dark Matter and the Dinosaurs: The Astounding Interconnectedness of the Universe.* New York: Ecco, 2015.

Randall, Lisa. *Warped Passages: Unraveling the Mysteries of the Universe's Hidden Dimensions.* New York: Ecco, 2005.

Robinson, Marilynne. *Gilead.* New York: Picador, 2020.

Schwarzschild, Karl. Karl Schwarzschild to Albert Einstein, December 22, 1915. In *The Collected Papers of Albert Einstein, Volume 8: The Berlin Years Correspondence 1914–1918* (English translation supplement), translated by Ann M. Hentschel, 163. Princeton, NJ: Princeton University Press, 1998.

Voltaire. *The Elements of Newton's Philosophy.* Translated by John Hanna. London: Routledge, 2019.

Wheeler, John Archibald, and Kenneth Ford. *Geons, Black Holes, and Quantum Foam, A Life in Physics.* New York: Norton, 2000.

Wigner, Eugene. "The Unreasonable Effectiveness of Mathematics in the Natural Sciences." *Communications in Pure and Applied Mathematics* 13, no. 1 (February 1960): 1–14.

Zenith, Richard. *Pessoa: A Biography.* New York: Norton, 2021.